Advanced Catalysis in Hydrogen Production from Formic Acid and Methanol

Advanced Catalysis in Hydrogen Production from Formic Acid and Methanol

Special Issue Editor

Dmitri A. Bulushev

MDPI • Basel • Beijing • Wuhan • Barcelona • Belgrade • Manchester • Tokyo • Cluj • Tianjin

Special Issue Editor
Dmitri A. Bulushev
Boreskov Institute of
Catalysis SB RAS
Russia

Editorial Office
MDPI
St. Alban-Anlage 66
4052 Basel, Switzerland

This is a reprint of articles from the Special Issue published online in the open access journal *Energies* (ISSN 1996-1073) (available at: https://www.mdpi.com/journal/energies/special_issues/Hydrogen_Production_Formic_Acid_Methanol).

For citation purposes, cite each article independently as indicated on the article page online and as indicated below:

LastName, A.A.; LastName, B.B.; LastName, C.C. Article Title. *Journal Name* **Year**, *Article Number*, Page Range.

ISBN 978-3-03936-380-3 (Pbk)
ISBN 978-3-03936-381-0 (PDF)

Contents

About the Special Issue Editor . . . . **vii**

Preface to "Advanced Catalysis in Hydrogen Production from Formic Acid and Methanol" . . **ix**

Fedor S. Golub, Sergey Beloshapkin, Artem V. Gusel'nikov, Vasily A. Bolotov, Valentin N. Parmon and Dmitri A. Bulushev
Boosting Hydrogen Production from Formic Acid over Pd Catalysts by Deposition of N-Containing Precursors on the Carbon Support
Reprinted from: *Energies* **2019**, *12*, 3885, doi:10.3390/en12203885 . . . . **1**

Alina D. Nishchakova, Dmitri A. Bulushev, Olga A. Stonkus, Igor P. Asanov, Arcady V. Ishchenko, Alexander V. Okotrub and Lyubov G. Bulusheva
Effects of the Carbon Support Doping with Nitrogen for the Hydrogen Production from Formic Acid over Ni Catalysts
Reprinted from: *Energies* **2019**, *12*, 4111, doi:10.3390/en12214111 . . . . **15**

Olga Podyacheva, Alexander Lisitsyn, Lidiya Kibis, Andrei Boronin, Olga Stonkus, Vladimir Zaikovskii, Arina Suboch, Vladimir Sobolev and Valentin Parmon
Nitrogen Doped Carbon Nanotubes and Nanofibers for Green Hydrogen Production:
Similarities in the Nature of Nitrogen Species, Metal–Nitrogen Interaction,
and Catalytic Properties
Reprinted from: *Energies* **2019**, *12*, 3976, doi:10.3390/en12203976 . . . . **25**

Alexey Pechenkin, Sukhe Badmaev, Vladimir Belyaev and Vladimir Sobyanin
Production of Hydrogen-Rich Gas by Formic Acid Decomposition over
$CuO\text{-}CeO_2/\gamma\text{-}Al_2O_3$ Catalyst
Reprinted from: *Energies* **2019**, *12*, 3577, doi:10.3390/en12183577 . . . . **39**

Vladimir Sobolev, Igor Asanov and Konstantin Koltunov
The Role of Support in Formic Acid Decomposition on Gold Catalysts
Reprinted from: *Energies* **2019**, *12*, 4198, doi:10.3390/en12214198 . . . . **49**

Miriam Navlani-García, David Salinas-Torres and Diego Cazorla-Amorós
Hydrogen Production from Formic Acid Attained by Bimetallic Heterogeneous
PdAg Catalytic Systems
Reprinted from: *Energies* **2019**, *12*, 4027, doi:10.3390/en12214027 . . . . **57**

Panagiota Stathi, Maria Solakidou, Maria Louloudi and Yiannis Deligiannakis
From Homogeneous to Heterogenized Molecular Catalysts for H_2 Production by Formic Acid
Dehydrogenation: Mechanistic Aspects, Role of Additives, and Co-Catalysts
Reprinted from: *Energies* **2020**, *13*, 733, doi:10.3390/en13030733 . . . . **85**

About the Special Issue Editor

Dmitri A. Bulushev graduated from the Novosibirsk State University (Russia) in 1983 and received his PhD degree in chemistry (chemical kinetics and catalysis) from the Boreskov Institute of Catalysis (Russia) in 1991. From 1995 to 2007, he was a researcher at the University of Ghent (Belgium) and EPFL (Switzerland), then a senior research fellow at the University of Limerick (Ireland). Since 2014, he has been working as a senior researcher at the Boreskov Institute of Catalysis. He has published more than 70 peer-reviewed papers, which have received more than 2700 citations. Dmitri has presented his results orally at top catalysis congresses (International Congress on Catalysis and EuropaCat). He has completed the supervision of two PhD students. Dmitri is a reviewer for more than 50 international journals and 10 science-funding organizations. Since 2019, he has been a member of the Editorial Board of the MDPI journal Energies. His research interests include hydrogen production, conversion of biomass to valuable chemicals, factors determining the activity and selectivity of catalysts, and the application of novel carbon materials as catalyst supports.

Preface to "Advanced Catalysis in Hydrogen Production from Formic Acid and Methanol"

Studies of catalytic decomposition of formic acid and methanol have more than 100 years of history, contributing significantly to the science and applications of catalysis. These investigations are related to the name of Paul Sabatier, who was awarded the Nobel Prize in Chemistry in 1912. Formic acid and methanol contain a sufficient amount of hydrogen that can be liberated at mild conditions using catalysis. Formic acid and methanol are considered as liquid organic hydrogen carriers (LOHCs) and can be produced using renewable methods from biomass or by CO_2 hydrogenation. Therefore, the studies in this field recently have become focused on the development of efficient catalysts for hydrogen production from these compounds. The aim of the Special Issue "Advanced Catalysis in Hydrogen Production from Formic Acid and Methanol" was to discuss the field of hydrogen production by the catalytic decomposition (or steam-reforming) of formic acid and methanol. The questions related to preparation and characterization of efficient homogeneous or heterogeneous catalysts, reaction mechanisms and kinetics, and reactor systems engineering have been discussed in this Issue. The researchers have submitted the original results of their theoretical and experimental work.

Dmitri A. Bulushev
Special Issue Editor

Article

Boosting Hydrogen Production from Formic Acid over Pd Catalysts by Deposition of N-Containing Precursors on the Carbon Support

Fedor S. Golub [1,2], Sergey Beloshapkin [3], Artem V. Gusel'nikov [4], Vasily A. Bolotov [1], Valentin N. Parmon [1,2] and Dmitri A. Bulushev [1,2,*]

[1] Laboratory of Catalytic Methods of Solar Energy Transformation, Boreskov Institute of Catalysis, SB RAS, 630090 Novosibirsk, Russia; fedorglb@gmail.com (F.S.G.); bolotov@catalysis.ru (V.A.B.); parmon@catalysis.ru (V.N.P.)
[2] Department of Natural Sciences, Novosibirsk State University, 630090 Novosibirsk, Russia
[3] Bernal Institute, University of Limerick, V94 T9PX Limerick, Ireland; serguei.belochapkine@ul.ie
[4] Laboratory of Physico-Chemistry of Nanomaterials, Nikolaev Institute of Inorganic Chemistry, SB RAS, 630090 Novosibirsk, Russia; artemg@ngs.ru
* Correspondence: dmitri.bulushev@catalysis.ru

Received: 18 September 2019; Accepted: 11 October 2019; Published: 14 October 2019

Abstract: Formic acid is a promising liquid organic hydrogen carrier (LOHC) since it has relatively high hydrogen content (4.4 wt%), low inflammability, low toxicity and can be obtained from biomass or from CO_2. The aim of the present research was the creation of efficient 1 wt% Pd catalysts supported on mesoporous graphitic carbon (Sibunit) for the hydrogen production from gas-phase formic acid. For this purpose, the carbon support was modified by pyrolysis of deposited precursors containing pyridinic nitrogen such as melamine (Mel), 2,2′-bipyridine (Bpy) or 1,10-phenanthroline (Phen) at 673 K. The following activity trend of the catalysts Pd/Mel/C > Pd/C ~ Pd/Bpy/C > Pd/Phen/C was obtained. The activity of the Pd/Mel/C catalyst was by a factor of 4 higher than the activity of the Pd/C catalyst at about 373 K and the apparent activation energy was significantly lower than those for the other catalysts (32 vs. 42–46 kJ/mol). The high activity of the melamine-based samples was explained by a high dispersion of Pd nanoparticles (~2 nm, HRTEM) and their strong electron-deficient character (XPS) provided by interaction of Pd with pyridinic nitrogen species of the support. The presented results can be used for the development of supported Pd catalysts for hydrogen production from different liquid organic hydrogen carriers.

Keywords: hydrogen production; formic acid decomposition; Pd/C; melamine; g-C_3N_4; bipyridine; phenanthroline; N-doped carbon

---

1. Introduction

According to research performed for the US Department of Energy [1], it is expected that despite growth in non-fossil fuel consumption will be greater than that of fossil fuels, the fossil fuels will still account for 78% of energy use in 2040. The resulting problems are obvious. Among them are reducing non-renewable sources of energy and high CO_2 emission, which are suspected to excite global warming with all related problems. For this reason, interest in eco-friendly and renewable sources like hydrogen, water, solar or wind energy grows every year.

Hydrogen is a perspective fuel due to possibility of its applications including fuel cells. The main problems with the hydrogen use are its storage and transportation. The various methods based on physical or chemical storage have been developed to solve these problems. The physical methods assume storage of molecular hydrogen in vessels at high pressure and low temperatures or hydrogen adsorption on high surface area materials like porous carbon. However, both these methods lead to large energy losses; therefore, such storage may appear inefficient. The chemical method implies hydrogen storage in a chemical bounded form, in molecules with high hydrogen content. This hydrogen can be efficiently recovered by a catalytic or non-catalytic reaction.

At the present time, formic acid is considered as an efficient liquid organic hydrogen carrier (LOHC) [2]. It has sufficiently good physical properties such as stability under standard conditions, low inflammability, low toxicity (compared to methanol, ammonia or hydrazine, for example), and relatively high hydrogen content (4.4 wt%). It is important to note, that the density of gravimetric energy of formic acid is seven times higher than that of the currently used lithium-ion batteries [3]. In addition, formic acid can be obtained from biomass [4,5] or from CO_2 and renewable H_2 [6].

A lot of studies related to reversible chemical hydrogen storage [7,8] provide access to potentially inexpensive, highly efficient and rechargeable hydrogen fuel cells. It is safer to use an easily decomposable hydrogen-containing material such as formic acid instead of storage and transportation of molecular hydrogen.

The catalytic formic acid decomposition reaction can be performed either in liquid phase or in gas phase. The liquid-phase decomposition is only rarely performed continuously and the determined reaction rates in this case often correspond to the initial rates and not to the steady-state. Gas-phase formic acid decomposition follows two possible pathways: through dehydrogenation to form H_2 and CO_2 (Equation (1)) and through dehydration to form CO and H_2O (Equation (2)):

$$HCOOH_{(gas)} \rightarrow H_2 + CO_2, \ \Delta_r H^o_{298} = -14.7 \ \frac{kJ}{mol} \tag{1}$$

$$HCOOH_{(gas)} \rightarrow H_2O_{(gas)} + CO, \ \Delta_r H^o_{298} = 26.5 \ \frac{kJ}{mol} \tag{2}$$

Depending on the catalysts and conditions the products of the Reactions (1) and (2) can transform into each other via the Water-Gas Shift or Reverse Water-Gas Shift Reactions. Only the catalysts providing very high selectivity towards Reaction (1) allow using formic acid and CO_2 cycle for a hydrogen storage system.

Catalysts containing no Pt-group metals [9,10] do not show promising properties as the catalysts with precious metals. Catalysts consisting of supported highly dispersed Pd particles [2,11,12] or even single Pd atoms [13–15] are often considered as the most efficient for the hydrogen production from formic acid. High dispersion of Pd implies high sensitivity of the metal properties to the support material. Flat graphitic carbon surfaces normally only weakly interact with Pd leading to agglomeration of metal and catalyst deactivation during the reaction. The approach to solving this problem is modifying the carbon surface by using other chemical elements. Modifying the surface by electron-rich nitrogen atoms increases the basicity of the support [16,17], sometimes leads to a higher dispersion of supported metal [10,17] and may change the nature of the active site. Recently, we have shown [15] for the gas phase formic acid decomposition, while Bi et al. [18] have reported for the liquid phase one that the pyridinic nitrogen sites present on the surface of carbon support can strongly interact with Pd and change its electronic state providing high activity in the reaction.

Introducing of pyridinic nitrogen on the surface of carbon support can be done by deposition of substances with pyridinic nitrogen [19–22]. This approach provides some advantages, as compared to direct synthesis of a CN material by chemical vapor deposition (CVD). The main advantage is a possibility to reach a high yield of N-containing material without a catalyst normally needed for the direct synthesis. In the latter case, the productivity of the catalyst is limited due to the blockage of the catalyst used for the CN material growth by carbon and complete deactivation. Therefore, the goal of the present work was the creation of efficient Pd catalysts for the hydrogen production from formic acid based on the N-doped carbon supports obtained by deposition of melamine, 1,10-phenanthroline and 2,2′-bipyridine on a mesoporous graphitic carbon. The choice of these N-containing precursors is explained by the presence of neighboring nitrogen atoms in their molecules. As we have shown earlier [14,15,23] the structure of the active site on the N-doped carbon may involve a Pd atom attached to a pair of pyridinic nitrogen atoms in the armchair position on the graphene edge. This armchair unit is present in the initial molecules of 2,2′-bipyridine and 1,10-phenanthroline and could be created during pyrolysis of melamine supported on carbon.

For comparisons a graphitic carbon nitride (g-C_3N_4) was also used as a support since it also possesses pyridinic nitrogen sites. Lee et al. [24] showed that catalysts comprising of Pd supported on graphitic carbon nitride (g-C_3N_4) were promising for liquid phase formic acid decomposition. Additionally, Oh [25] demonstrated that this type of catalysts may produce hydrogen from liquid formic acid continuously at least within 3 h, and this result can be used for development of fuel cells.

2. Materials and Methods

2.1. Synthesis of Supports

2.1.1. g-C_3N_4

g-C_3N_4 was prepared according to the procedure described by Tahir et al. [26]. This was chosen because of the reported high BET surface area of the obtained material. Nitric acid (180 mL, 0.2 M, NAK Azot, Novomoskovsk, Russia) was added to a mixture of melamine (3 g, Sigma-Aldrich, St. Louis, MO, USA) and ethylene glycol (60 mL, Acros Organics, Waltham, MA, USA) and stirred for 10 min until a white powder was precipitated. Then, the precipitate was washed with isopropyl alcohol and dried at 363 K. This sample was called Inter-C_xN_y. To obtain g-C_3N_4, the Inter-C_xN_y sample was heated in air in an already heated muffle furnace at 673 K for 15 min. Other heating modes were also used. It should be mentioned that melamine is inexpensive and low toxic substance [27].

2.1.2. Mel/C

The used approach of doping the carbon material by substances containing pyridinic nitrogen is shown in Figure 1. Melamine (0.75 g) was mixed with isopropyl alcohol (about 100 mL) at 343 K under constant stirring. Then, the mixture was added to the mesoporous graphitic carbon—Sibunit (1.5 g, Sibunit #6, Omsk, Russia). The obtained material was kept for 10 min while isopropyl alcohol was evaporating. Then, the sample was dried at 357 K and pyrolyzed in the muffle furnace (673 K, 1 K/min) in air for 2 h. This catalyst support was called Mel/C (muf). The used Sibunit carbon is known to possess high surface area and high mechanical strength [28].

Another preparation method of the support was heating of a mechanical mixture of melamine and Sibunit (1:2) under microwave (MW) radiation in air. Microwave treatment of the mixture was carried out using a specially developed equipment (operating frequency 2470 ± 10 MHz, power up to 1 kW) based on a rectangular single-mode cavity with high Q-factor (~6000), which is excited by TE_{102} oscillation mode, described in [29]. Processing MW conditions were the following: T = 673 K; heating rate −40 K/s and duration time −140 s. The temperature in the reactor was measured using an optical pyrometer (Raytek, Santa Cruz, CA, USA). The catalyst support obtained by this method was called Mel/C (MW).

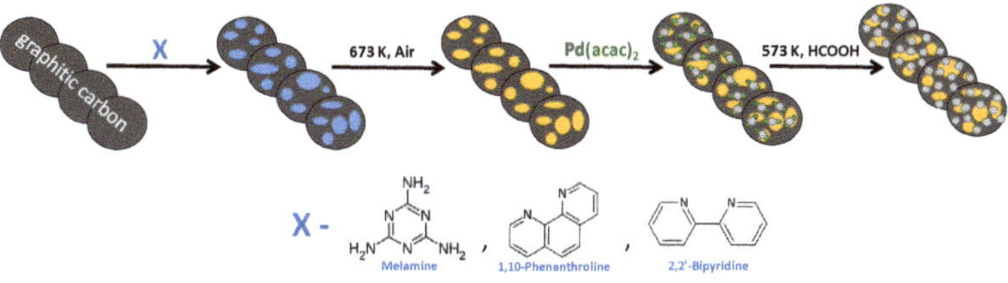

Figure 1. A scheme of synthesis of the catalysts based on graphitic carbon (Sibunit).

2.1.3. Phen/C and Bpy/C

Phenanthroline or bipyridine (Sigma-Aldrich, St. Louis, MO, USA) was dissolved in isopropyl alcohol (1 mL). The used precursor charge was selected in such a way that the deposited C_xN_y-material content in the synthesized sample was 14 wt% that was the same as for the Mel/C (muf) sample after the pyrolysis. Then, the prepared solution was added to the carbon sample and dried at 358 K. The'pyrolysis was performed in the muffle furnace (673 K, 1 K /min, 2 h) in air.

2.2. Synthesis of Pd Catalysts

A preparation of 1 wt% Pd catalysts was performed by impregnation of the synthesized supports with a Pd acetylacetonate (Sigma-Aldrich, St. Louis, MO, USA) solution in acetone (EKOS-1, Moscow, Russia). This mixture was kept for 5 min with constant stirring followed by drying at 343 K for 1 h (Figure 1).

A preparation of a (Pd+Phen)/C catalyst was different. First, phenanthroline and $Pd(acac)_2$/acetone solution were mixed in attempt to obtain a complex of Pd with phenanthroline. Then this mixture was added to the Sibunit carbon. Subsequent steps were the same as above.

2.3. Characterization

Textural samples characteristics were determined using nitrogen adsorption-desorption on an ASAP 2400 device (Micrometrics, Norcross, GA, USA) at 77 K. The samples were pretreated in a vacuum chamber at 423 K for 1 h. The calculation of the surface area was carried out using the Brunauer-Emmett-Teller (BET) method. A CHN-analysis was performed on a Vario EL Cube element analyzer (Elementar, Langenselbold, Germany) with a thermal conductivity detector.

X-ray Diffraction (XRD) patterns were recorded using Thermo X'tra (Waltham, MA, USA) and Bruker D8 Advance (Billerica, MA, USA) diffractometers with monochromatic $CuK\alpha$ radiation (λ = 1.5418 Å). On the patterns, the 2θ region lower than 15° is not shown since the used sample holder contributed there. To measure the size of Pd particles, a high-resolution transmission electron microscopy (HRTEM) study was performed using a JEM-2100F unit (JEOL, Akishima, Japan) with the acceleration voltage of 200 kV. Determination of electronic state and surface concentrations of Pd, N, O and C was done via an AXIS Ultra DLD photoelectron spectrometer (Kratos, Manchester, UK) with monochromatic $AlK\alpha$ radiation (1486.6 eV). C1s at 284.8 eV was chosen as the standard for energy calibration. To prevent charging of the samples the charge neutralizer was always switched on during the measurements.

2.4. Catalytic Measurements

Catalytic measurements were performed in a fixed-bed glass reactor attached to a catalytic flow set-up presented in Figure S1 (Supplementary Materials). Carrier gas (Ar) passed through a glass flask filled with liquid formic acid to saturate it with formic acid vapor. Then, the flow was additionally diluted with Ar until the formic acid concentration reached the value of 2.5 vol%. The total volumetric flow rate was constant and equal to 67 mL/min in all experiments. The temperature in the reactor was maintained with a furnace in the range from 323 to 673 K. The catalyst weight used in experiments corresponded to 10–16 mg. The products were analyzed by a gas chromatograph (Chromos GC-1000, Moscow, Russia) equipped with two thermal conductivity detectors and packed columns filled with CaA molecular sieves and HayeSep-Q porous polymer. For separation of the products, heating from 313 K to 483 K was performed within 10 min and then the temperature was kept stable for 20 min.

Before the measurements, the catalyst was treated with the same formic acid/Ar flow at 573 K for 20 min to decompose supported $Pd(acac)_2$ (decomposition temperature −478 K) and stabilize the catalyst (Figure 1). After cooling the catalyst in the formic acid/Ar flow, the measurements were started at 353 K and then performed with 15–20 K steps. The temperature was kept for 20 min at each step to reach the steady-state. Specific reaction rates (W) were determined basing on the Pd mass in the catalyst. They were calculated at low formic acid conversions using a formula:

$$W = \frac{X_{FA} \cdot C_{FA} \cdot v_f \cdot N_A}{22,400 \cdot m_{Pd}}, \quad \left(\frac{molecule}{s \cdot g_{Pd}}\right) \tag{3}$$

X_{FA}—formic acid conversion, v_f—total flow rate (ml/s), C_{FA}—formic acid concentration, m_{Pd}—mass of Pd in the sample (g), N_A—the Avogadro constant (6.02×10^{23} molecule/mol), 22,400—volume of one mole of gas at 273 K and 1 atm (ml/mol).

3. Results and Discussion

3.1. Characterization of the Supports

BET surface area values for the catalyst supports are shown in Table 1. They vary in a wide range from 8 m^2/g to 348 m^2/g for the g-C_3N_4 and Sibunit support, respectively (Table 1). Unfortunately, the used synthesis procedure of the g-C_3N_4 material did not provide as high surface area as that reported by Tahir et al. [26] (290 m^2/g) for unknown reasons. We used different heating modes for the intermediate, but it did not allow increasing the surface area of the obtained g-C_3N_4 material. For example, heating of the sample at 673 K for 2 h gave even a lower surface area (6.5 m^2/g).

Table 1. Characteristics of the catalysts, supports and some kinetic data.

Sample	Support	Support Surface Area, m^2/g	Mean Particle Size, nm (TEM)	E_a, kJ/mol	H_2 Selectivity (50 % conv.) %
Pd/g-C_3N_4	g-C_3N_4	8.1	2.6 ± 0.3	39 ± 1	>98
Pd/Mel/C (muf)	Mel/C (muf)	86.5	2.0 ± 0.3	32 ± 1	>98
Pd/Mel/C (MW)	Mel/C (MW)	42.0	2.3 ± 0.5	32 ± 1	95.7
Pd/Phen/C	Phen/C	301	–	43 ± 1	90.1
(Pd + Phen)/C	C	348	–	42 ± 1	92.1
Pd/Bpy/C	Bpy/C	228	–	45 ± 1	94.8
Pd/C	C	348	2.3 ± 0.3	46 ± 2	93.4

Doping of the Sibunit surface with substances containing pyridinic nitrogen leads to a decrease of the surface area owing to occupation of the pores by the nitrogen precursor. The decrease is the strongest when melamine is used. However, the surface areas of the melamine-based supports were still high as compared to that of g-C_3N_4 and reached 86.5 and 42 m^2/g.

X-ray diffraction patterns of the g-C_3N_4, Mel/C, Sibunit supports and the intermediate for the carbon nitride synthesis (Inter-C_xN_y) are shown in Figure 2. The pattern for the intermediate shows a lot of sharp peaks, differs from the patterns presented in literature for melamine [26,30,31] and melem [32], but is very close to the pattern of melaminium nitrate [30]. Hence, we assign the Inter-C_xN_y intermediate to melaminium nitrate. The multiple peaks of this compound disappear after pyrolysis at 673 K. The pattern for the obtained g-C_3N_4 sample shows only one intensive peak at 27.1° in the studied region. This peak is characteristic interplanar stacking peak for g-C_3N_4 (JCPDS 87-1526) [33]. It indicates the presence of the ordered structure of layers of heptazine which is a monomer of g-C_3N_4 (002 plane). Interesting that the formation of g-C_3N_4 from melaminium nitrate takes place at the temperature lower than those from melamine reported in literature (823 K) [31,33].

Diffraction patterns of the Sibunit containing samples show diffraction peaks at 25.5° and 43.1° for the (002) and (100) planes of graphite (JCPDS 41-1487), respectively. The structure of the obtained g-C_3N_4 is denser, as the stacking distance for the carbon nitride is by 0.2 Å smaller than that for the Sibunit carbon.

Comparing the XRD patterns for the N-containing samples with the original Sibunit sample, we conclude that g-C_3N_4 cannot be determined using XRD because of the low carbon nitride content (12–14%) in the Mel/C sample and close positions of the g-C_3N_4 and Sibunit characteristic peaks ((002)

plane). However, an important result is that multiple characteristic peaks of the melamine, melem or possible intermediates are absent (Figure 2). Evidently, that melamine or intermediates completely converted to supported carbon nitride or to another type of a CN containing material.

The $^{C}/_{N}$ atomic ratio was determined for our g-C$_3$N$_4$ support by CHN-analysis and was found to be equal to 0.63. It indicates higher nitrogen content in the sample as compared to stoichiometric g-C$_3$N$_4$ (0.75). Such low ratios were reported for carbon nitrides earlier [31,34] and could be explained by the relatively low pyrolysis temperature, which we used.

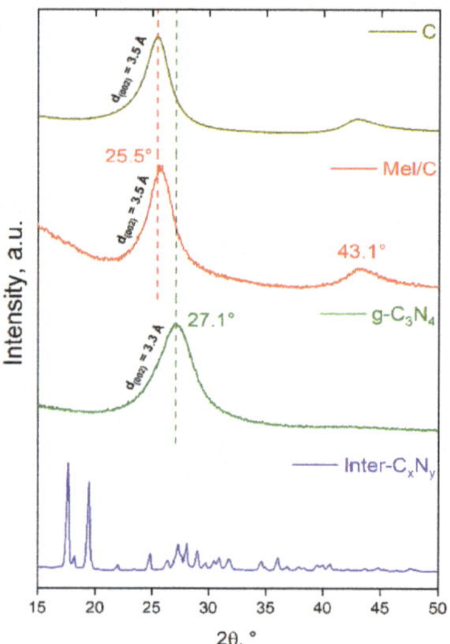

Figure 2. XRD patterns of the supports and intermediate of the g-C$_3$N$_4$ synthesis.

3.2. Catalytic Activity and H$_2$ Selectivity

To elucidate the support effect, using the obtained dependences between the formic acid conversion and temperature, we calculated the reaction rates at low conversions and plotted them as the Arrhenius plots (Figure 3). It is seen that both Pd/Mel/C samples convert formic acid better than the other samples (Pd/C and Pd/g-C$_3$N$_4$). The sample synthesized in the muffle furnace is slightly more active than the sample prepared using the microwave synthesis. However, it is worth noting that the synthesis time in the muffle furnace was 2 h, while that for the microwave synthesis was only 140 s. The difference in the rates between the melamine based Pd samples and the Pd/C sample is significant. The rates differ by a factor of 4 at about 373 K. In contrast, the Pd/g-C$_3$N$_4$ sample showed the activity even lower than that for the Pd/C catalyst.

Analyzing the data presented in Figure 4, we can estimate the N-precursor effect on the catalytic activity. The catalysts prepared with bipyridine and phenanthroline show lower activities than that for the melamine-based catalyst, and they are not higher than the one of the Pd/C catalysts. These results are consistent with the data for the apparent activation energies (E$_a$, Table 1). The most active Pd/Mel/C catalysts show the smallest activation energy (32 kJ/mol), while the less active catalysts demonstrate higher apparent activation energies (42–46 kJ/mol). The E$_a$ value for the Pd/g-C$_3$N$_4$ catalyst is intermediate (39 kJ/mol). The difference in the apparent activation energies is significant and should be related to different properties of active sites of the catalyst. The obtained E$_a$ values for the Pd/Mel/C catalyst are among the smallest values for formic acid decomposition over Pd catalysts in liquid and gas phase.

Important features necessary for efficient hydrogen production from formic acid are selectivity and stability. The Pd/Mel/C (muf) catalyst showed the highest selectivity (>98%) at 50% conversion (528 K) while the phenanthroline based samples showed the lowest selectivities (90–92%).

The stability of the most active Pd/Mel/C (muf) catalyst was tested for 5 h (Figure S2). In this experiment, the catalyst was not stabilized by pretreatment in the formic acid flow at 573 K as was done before other experiments. However, no significant changes of the formic acid conversion on time-on-stream was observed, and it is possible to conclude that this catalyst possesses sufficient stability to produce hydrogen from gas phase formic acid.

Hence, among the studied samples, the melamine based Pd catalysts showed the highest activity. We compared the activities of these samples with the activities of 1 wt% Ru, Pt and Au samples on the N-doped and N-free carbon supports (Table S1). This comparison showed that the metal mass-based activity of the Pd/Mel/C (muf) catalyst is slightly lower than that of the Pt/N-C catalyst, but it is higher than the activities of the other catalysts. The apparent activation energy for the melamine based Pd sample was the lowest.

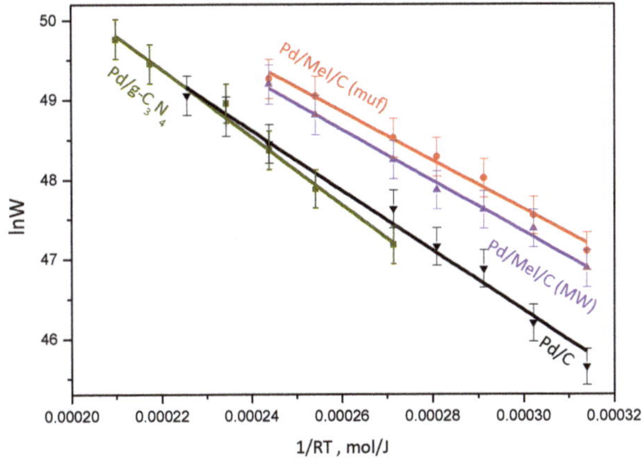

Figure 3. Arrhenius plots for formic acid decomposition showing the support effect.

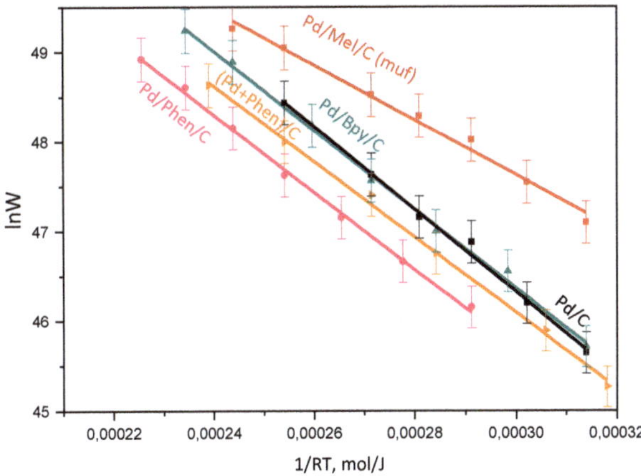

Figure 4. Arrhenius plots for formic acid decomposition showing the N-precursor effect.

3.3. Characterization of the Pd Catalysts

To understand the reasons of the differences in the activities of the catalysts (Figures 3 and 4) the Pd samples after the reaction were studied using HRTEM and X-ray Photoelectron spectroscopy (XPS). Figure 5 shows HRTEM images. Particle size distributions are presented in Figure S3. They are quite narrow for all studied samples indicating that the particles are rather uniform. Mean particle size data are presented in Table 1. The most active Pd/Mel/C (muf) catalyst demonstrated the smallest mean particle size (2.0 nm). An XRD study of this sample was not sensitive enough to determine Pd in this sample confirming its low content and high dispersion. The highest mean particle size was obtained for the Pd/g-C_3N_4 sample, which corresponds to 2.6 nm. This could be related to the lowest BET surface area of the support used (Table 1). However, the mean Pd particle sizes for other two catalysts did not differ much. Generally, as it is seen in Table 1, the deposition of nitrogen precursors over the carbon surface did not provide a significant change in Pd dispersion.

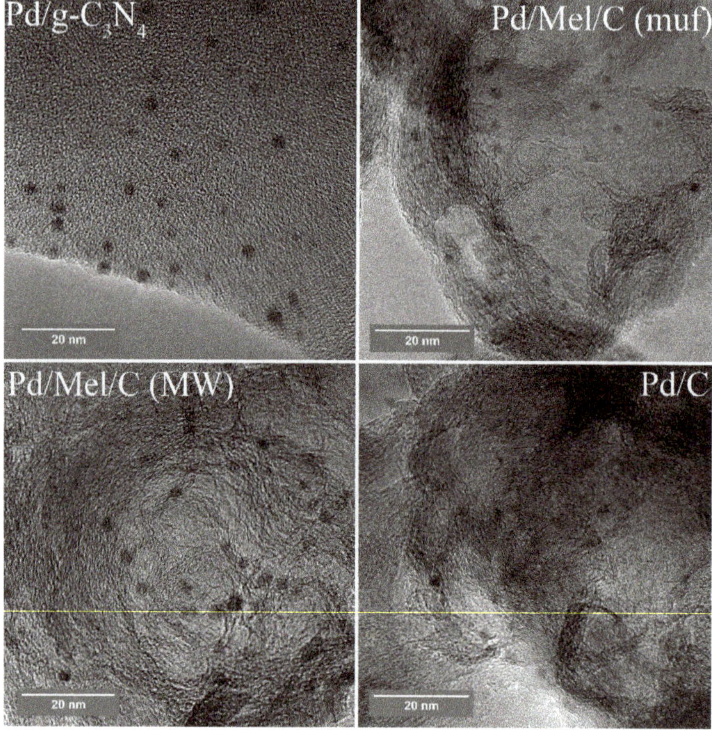

Figure 5. HRTEM images of some catalysts after the reaction.

A C1s XP spectrum for the Pd/g-C_3N_4 catalyst after the reaction is shown in Figure S4. It consists of two main peaks at 288.3 and 284.8 eV assigned to pyridinic (C-N=C) and adventitious carbon, respectively. The presence of the first peak confirms the formation of g-C_3N_4, as it is typical for this compound [26,33]. The C1s spectra for other catalysts did not differ much from each other and, hence, are not demonstrated.

N1s XP spectra for all catalysts after the reaction are shown in Figure 6. The main lines are observed at 398.8 eV assigned to pyridinic nitrogen (C-N=C) and at 400.2 eV assigned to tertiary nitrogen (N(C)$_3$) like those in g-C_3N_4 [33]. A shift of the N1s pyridinic line towards higher binding energies by 0.3 eV is observed for the Pd/Mel/C (MW) catalyst. This may indicate a stronger interaction of these species in the surface layer of this sample than in the case of the Pd/Mel/C (muf) and Pd/g-C$_3$N$_4$ samples.

Table 2 shows the total surface N concentration determined by XPS (N$_{tot}$). The highest concentration was obtained for the Pd/g-C$_3$N$_4$ catalyst (46%) followed by the Pd/Mel/C (MW) (8.9 at %) and Pd/Mel/C (muf) (4.5 at %) catalysts. The spectra of the melamine-based samples demonstrate the presence of both N1s lines, but the ratio of pyridinic species is decreased as compared to the Pd/g-C$_3$N$_4$ sample (Table 2). The content of pyridinic N species in the Pd/Phen/C sample is negligible and Pd/Bpy/C and (Pd+Phen)/C samples do not contain pyridinic nitrogen at all. This shows that the pyridinic species have been converted either to tertiary N species or to gas products during the pyrolysis. The difference in the spectra of the melamine based and phenanthroline/bipyridine based samples can be explained at least partly by a much lower content of nitrogen in the bipyridine and phenanthroline precursors as compared to melamine (Figure 1). Generally, the state of the deposited N-containing layer in the supported samples differs from that in the g-C$_3$N$_4$ sample. This can affect the state of supported Pd and provide the observed difference in catalytic properties (Figures 3 and 4). Therefore, the electronic state of Pd was also studied by XPS.

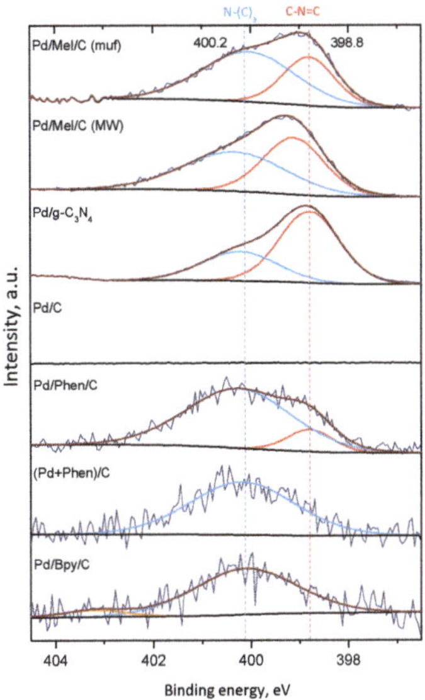

Figure 6. XP spectra of the N1s region for the catalysts after the reaction.

Table 2. Surface concentrations of N species in the catalysts after the reaction (XPS data).

Sample	Pd/g-C$_3$N$_4$	Pd/Mel/C (muf)	Pd/Mel/C (MW)	Pd/Phen/C	(Pd+Phen)/C	Pd/Bpy/C	Pd/C
N$_{tot}$, at %	46.0	4.5	8.9	1.35	0.57	0.48	0
$\frac{C\text{-}N\text{=}C}{N_{tot}}$	0.66	0.35	0.46	0.14	0	0	–

Pd3d XP spectra are shown in Figure 7. The spectra consist of two components: one with Pd3d^5/$_2$ at 335.0–336.4 eV (Pd3d^3/$_2$ at 340.3–341.7 eV) and another with Pd3d^5/$_2$ at 336.9–338.5 eV (Pd3d^3/$_2$ at 342.2–343.7 eV). Taking into account the HRTEM data (Figure 5) similar for all samples, we attribute the first component to metallic Pd (Pd^0), and the second component—to oxidized Pd (Pd^{2+}). The latter can exist as a surface oxide over Pd particles. It is formed due to oxidation of the metallic Pd surface by atmospheric oxygen. Alternatively, some Pd^{2+} species may exist in atomically dispersed state strongly

interacting with nitrogen, oxygen or carbon defects of the support [13–15,35,36]. This Pd could not be observed by the HRTEM equipment used in this work.

The Pd3d$_{5/2}$ line at 335.0 eV is observed for the Pd/g-C$_3$N$_4$ catalyst, which is close to the Pd state in the unsupported bulk metal, for example, in Pd foil. The closeness of the electronic state of Pd in the sample to the Pd metal may indicate that there is no strong interaction of this Pd with the support. This can be due to small support surface area and low concentration of support sites, which can strongly interact with Pd. The lowest value of the Pd3d$^{5/2}$ binding energy among the Sibunit-containing catalysts is observed for the Pd/C catalyst and corresponds to 335.8 eV. This Pd state is usually attributed to small metal particles interacting with the carbon support [13,35,37]. The Pd3d$^{5/2}$ lines for the Pd/Mel/C (muf) and Pd/Mel/C (MW) catalysts are strongly shifted to higher binding energies (336.4 and 336.3 eV, respectively) as compared to that for the Pd/C sample. This cannot be assigned to sample charging and indicates an electron-deficient state of Pd in these samples due to a strong interaction with the support. These catalysts showed the highest activity in the formic acid decomposition reaction and the lowest values of apparent activation energies (Table 1). For the phenanthroline and bipyridine-based samples, the Pd3d$^{5/2}$ line positions (336.0 eV) are close to that for the Pd/C sample. The catalytic properties of these systems are also close (Figure 4).

It is an interesting question whether the Pd3d$^{5/2}$ line in the 336.3–336.4 eV region for the melamine-based samples can be assigned to Pd oxide (PdO). We believe that it cannot, because the Pd3d$^{5/2}$ binding energy for the PdO is normally located at a higher binding energy (336.8 eV) [38,39] like those observed for the Pd/g-C$_3$N$_4$ and Pd/C samples. However, we will perform soon the necessary experiments allowing to discriminate whether the high binding energies for the Pd3d$_{5/2}$ line (336.3–336.4 eV) should be assigned to Pd metal or to Pd oxide by performing XPS measurements after the pretreatment of the samples in H$_2$ in the pretreatment chamber of the XP spectrometer, as it was done in some of our previous works [10,13,14]. This pretreatment easily transforms the Pd oxide into metallic Pd.

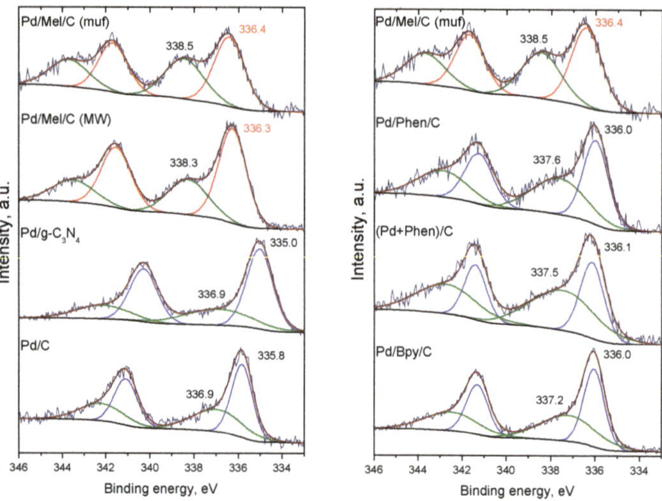

Figure 7. Pd3d XP spectra for the catalysts after the reaction.

Hence, utilization of HRTEM and XPS for the catalysts allowed explaining the effects observed for the catalytic reaction. The obtained data demonstrated that when pyridinic nitrogen exists on the surface, Pd becomes electron-deficient and the highest catalytic activity was obtained. At the same time, the presence of only tertiary nitrogen species does not positively affect the reaction.

4. Conclusions

In order to obtain a high surface area N-doped support, N-containing precursors (melamine, phenanthroline or bipyridine) were deposited on the mesoporous graphitic carbon material (Sibunit) and pyrolysis at 673 K was performed. Melamine showed itself as the most promising N-precursor. The high activity of the melamine based catalysts in the hydrogen production from formic acid can be explained by a strong interaction between highly dispersed Pd nanoparticles (~2 nm) and excess of pyridinic nitrogen of the support that can bind Pd particles and provide a shift of electron density from the Pd to the support, thus, transforming Pd into an electron-deficient state. This state demonstrates a significantly enhanced catalytic activity (by a factor of 4 at 373 K), high stability in the reaction (5 h) and the lowest apparent activation energy (32 kJ/mol) as compared to the state of Pd in the other carbon supported catalysts (42–46 kJ/mol).

The used approach for synthesis of the N-doped carbon with melamine is much simpler and, if necessary, it gives a higher yield of the N-doped carbon support as compared to the direct catalytic CVD approaches. Additionally, since the used carbon support (Sibunit) was specifically developed to have high surface area and high mechanical strength, our supported Pd catalysts possessed very much the same properties. Hence, the melamine based Pd catalysts are promising for the hydrogen production from formic acid and also, probably, from other liquid organic hydrogen carriers.

Supplementary Materials: The following are available online at http://www.mdpi.com/1996-1073/12/20/3885/s1, Figure S1: Scheme of a catalytic set-up, Figure S2: Stability test for the most active catalyst. Figure S3: Particle size distributions for some catalysts after the reaction, Figure S4: C1s XP spectrum of the Pd/g-C_3N_4 catalyst after the reaction.

Author Contributions: Synthesis, measurements and writing, F.S.G. and V.A.B.; characterization, S.B.; equipment creation, A.V.G.; funding security and content editing, V.N.P.; methodology, data analysis and writing, D.A.B.

Funding: The work was supported by the Russian Science Foundation (grant 17-73-30032).

Acknowledgments: The authors thank O.A. Bulavchenko for XRD measurements.

Conflicts of Interest: The authors declare no conflict of interest.

References

1. US Energy Information Administration. International Energy Outlook 2016. 2016; No. 0484. Available online: www.eia.gov/outlooks/ieo/pdf/0484(2016).pdf (accessed on 14 October 2019).

2. Zhong, H.; Iguchi, M.; Chatterjee, M.; Himeda, Y.; Xu, Q.; Kawanami, H. Formic Acid-Based Liquid Organic Hydrogen Carrier System with Heterogeneous Catalysts. *Adv. Sust. Syst.* **2018**, *2*, 1700161. [CrossRef]

3. Boddien, A.; Federsel, C.; Sponholz, P.; Mellmann, D.; Jackstell, R.; Junge, H.; Laurenczy, G.; Beller, M. Towards the Development of a Hydrogen Battery. *Energy Environ. Sci.* **2012**, *5*, 8907–8911. [CrossRef]

4. Bulushev, D.A.; Ross, J.R.H. Towards Sustainable Production of Formic Acid. *ChemSusChem* **2018**, *11*, 821–836. [CrossRef] [PubMed]

5. Reichert, J.; Brunner, B.; Jess, A.; Wasserscheid, P.; Albert, J. Biomass Oxidation to Formic Acid in Aqueous Media Using Polyoxometalate Catalysts—Boosting Fa Selectivity by in-Situ Extraction. *Energy Environ. Sci.* **2015**, *8*, 2985–2990. [CrossRef]

6. Bulushev, D.A.; Ross, J.R.H. Heterogeneous Catalysts for Hydrogenation of CO_2 and Bicarbonates to Formic Acid and Formates. *Catal. Rev.* **2018**, *60*, 566–593. [CrossRef]

7. Germain, J.; Hradil, J.; Fréchet, J.M.J.; Svec, F. High Surface Area Nanoporous Polymers for Reversible Hydrogen Storage. *Chem. Mater.* **2006**, *18*, 4430–4435. [CrossRef]

8. Hull, J.F.; Himeda, Y.; Wang, W.-H.; Hashiguchi, B.; Periana, R.; Szalda, D.J.; Muckerman, J.T.; Fujita, E. Reversible Hydrogen Storage Using CO_2 and a Proton-Switchable Iridium Catalyst in Aqueous Media under Mild Temperatures and Pressures. *Nat. Chem.* **2012**, *4*, 383–388. [CrossRef]

9. Koroteev, V.O.; Bulushev, D.A.; Chuvilin, A.L.; Okotrub, A.V.; Bulusheva, L.G. Nanometer-Sized MoS2 Clusters on Graphene Flakes for Catalytic Formic Acid Decomposition. *ACS Catal.* **2014**, *4*, 3950–3956. [CrossRef]

10. Bulushev, D.A.; Chuvilin, A.L.; Sobolev, V.I.; Stolyarova, S.G.; Shubin, Y.V.; Asanov, I.P.; Ishchenko, A.V.; Magnani, G.; Riccò, M.; Okotrub, A.V.; et al. Copper on Carbon Materials: Stabilization by Nitrogen Doping. *J. Mat. Chem. A* **2017**, *5*, 10574–10583. [CrossRef]

11. Wang, Q.; Tsumori, N.; Kitta, M.; Xu, Q. Fast Dehydrogenation of Formic Acid over Palladium Nanoparticles Immobilized in Nitrogen-Doped Hierarchically Porous Carbon. *ACS Catal.* **2018**, *8*, 12041–12045. [CrossRef]

12. Navlani-García, M.; Mori, K.; Salinas-Torres, D.; Kuwahara, Y.; Yamashita, H. New Approaches toward the Hydrogen Production from Formic Acid Dehydrogenation over Pd-Based Heterogeneous Catalysts. *Front. Mat.* **2019**, *6*. [CrossRef]

13. Podyacheva, O.; Bulushev, D.; Suboch, A.; Svintsitskiy, D.; Lisitsyn, A.; Modin, E.; Chuvilin, A.; Gerasimov, E.; Sobolev, V.; Parmon, V. Highly Stable Single-Atom Catalyst with Ionic Pd Active Sites Supported on N-Doped Carbon Nanotubes for Formic Acid Decomposition. *ChemSusChem* **2018**, *11*, 3724–3727. [CrossRef] [PubMed]

14. Zacharska, M.; Bulusheva, L.G.; Lisitsyn, A.S.; Beloshapkin, S.; Guo, Y.; Chuvilin, A.L.; Shlyakhova, E.V.; Podyacheva, O.Y.; Leahy, J.J.; Okotrub, A.V.; et al. Factors Influencing the Performance of Pd/C Catalysts in the Green Production of Hydrogen from Formic Acid. *ChemSusChem* **2017**, *10*, 720–730. [CrossRef] [PubMed]

15. Bulushev, D.A.; Zacharska, M.; Shlyakhova, E.V.; Chuvilin, A.L.; Guo, Y.; Beloshapkin, S.; Okotrub, A.V.; Bulusheva, L.G. Single Isolated Pd^{2+} Cations Supported on N-Doped Carbon as Active Sites for Hydrogen Production from Formic Acid Decomposition. *ACS Catal.* **2016**, *6*, 681–691. [CrossRef]

16. Arrigo, R.; Hävecker, M.; Wrabetz, S.; Blume, R.; Lerch, M.; McGregor, J.; Parrott, E.P.J.; Zeitler, J.A.; Gladden, L.F.; Knop-Gericke, A.; et al. Tuning the Acid/Base Properties of Nanocarbons by Functionalization Via Amination. *J. Am. Chem. Soc.* **2010**, *132*, 9616–9630. [CrossRef]

17. Bulushev, D.A.; Sobolev, V.I.; Pirutko, L.V.; Starostina, A.V.; Asanov, I.P.; Modin, E.; Chuvilin, A.L.; Gupta, N.; Okotrub, A.V.; Bulusheva, L.G. Hydrogen Production from Formic Acid over Au Catalysts Supported on Carbon: Comparison with Au Catalysts Supported on SiO_2 and Al_2O_3. *Catalysts* **2019**, *9*, 376. [CrossRef]

18. Bi, Q.-Y.; Lin, J.-D.; Liu, Y.-M.; He, H.-Y.; Huang, F.-Q.; Cao, Y. Dehydrogenation of Formic Acid at Room Temperature: Boosting Palladium Nanoparticle Efficiency by Coupling with Pyridinic-Nitrogen-Doped Carbon. *Angew. Chem. Int. Ed.* **2016**, *55*, 11849–11853. [CrossRef]

19. Li, Z.; Li, J.; Liu, J.; Zhao, Z.; Xia, C.; Li, F. Palladium Nanoparticles Supported on Nitrogen-Functionalized Active Carbon: A Stable and Highly Efficient Catalyst for the Selective Hydrogenation of Nitroarenes. *ChemCatChem* **2014**, *6*, 1333–1339. [CrossRef]

20. Gribov, E.N.; Kuznetsov, A.N.; Golovin, V.A.; Krasnikov, D.V.; Kuznetsov, V.L. Effect of Modification of Multi-Walled Carbon Nanotubes with Nitrogen-Containing Polymers on the Electrochemical Performance of Pt/CNT Catalysts in PEMFC. *Mater. Renew. Sustain. Energy* **2019**, *8*, 7. [CrossRef]

21. Salinas-Torres, D.; Navlani-García, M.; Mori, K.; Kuwahara, Y.; Yamashita, H. Nitrogen-Doped Carbon Materials as a Promising Platform toward the Efficient Catalysis for Hydrogen Generation. *Appl. Catal. A* **2018**, *571*, 25–41. [CrossRef]

22. Cai, J.; Bennici, S.; Shen, J.; Auroux, A. Influence of N Addition in Mesoporous Carbons Used as Supports of Pt, Pd and Ru for Toluene Hydrogenation and Iron Oxide for Benzene Oxidation. *React. Kinet. Mech. Catal.* **2015**, *115*, 263–282. [CrossRef]

23. Bulushev, D.A.; Zacharska, M.; Lisitsyn, A.S.; Podyacheva, O.Y.; Hage, F.S.; Ramasse, Q.M.; Bangert, U.; Bulusheva, L.G. Single Atoms of Pt-Group Metals Stabilized by N-Doped Carbon Nanofibers for Efficient Hydrogen Production from Formic Acid. *ACS Catal.* **2016**, *6*, 3442–3451. [CrossRef]

24. Lee, J.H.; Ryu, J.; Kim, J.Y.; Nam, S.W.; Han, J.H.; Lim, T.H.; Gautam, S.; Chae, K.H.; Yoon, C.W. Carbon Dioxide Mediated, Reversible Chemical Hydrogen Storage Using a Pd Nanocatalyst Supported on Mesoporous Graphitic Carbon Nitride. *J. Mater. Chem. A* **2014**, *2*, 9490–9495. [CrossRef]

25. Oh, T.H. A Formic Acid Hydrogen Generator Using Pd/C3N4 Catalyst for Mobile Proton Exchange Membrane Fuel Cell Systems. *Energy* **2016**, *112*, 679–685. [CrossRef]

26. Tahir, M.; Cao, C.; Butt, F.K.; Butt, S.; Idrees, F.; Ali, Z.; Aslam, I.; Tanveer, M.; Mahmood, A.; Mahmood, N. Large Scale Production of Novel G-C3N4 Micro Strings with High Surface Area and Versatile Photodegradation Ability. *CrystEngComm* **2014**, *16*, 1825–1830. [CrossRef]

27. Material Safety Data Sheet. Melamine. Available online: https://fscimage.fishersci.com/msds/96668.htm (accessed on 14 October 2019).

28. Fenelonov, V.B.; Likholobov, V.A.; Derevyankin, A.Y.; Mel'gunov, M.S. Porous Carbon Materials Prepared from C1–C3 Hydrocarbons. *Catal. Today* **1998**, *42*, 341–345. [CrossRef]

29. Chernousov, Y.D.; Shebolaev, I.V.; Ivannikov, V.I.; Ikryanov, I.M.; Bolotov, V.A.; Tanashev, Y.Y. An Apparatus for Performing Chemical Reactions under Microwave Heating of Reagents. *Instrum. Exp. Tech.* **2019**, *62*, 289–294. [CrossRef]

30. Vella-Zarb, L.; Braga, D.; Orpen, A.G.; Baisch, U. The Influence of Hydrogen Bonding on the Planar Arrangement of Melamine in Crystal Structures of Its Solvates, Cocrystals and Salts. *CrystEngComm* **2014**, *16*, 8147–8159. [CrossRef]

31. Papailias, I.; Giannakopoulou, T.; Todorova, N.; Demotikali, D.; Vaimakis, T.; Trapalis, C. Effect of Processing Temperature on Structure and Photocatalytic Properties of G-C3N4. *Appl. Surf. Sci.* **2015**, *358*, 278–286. [CrossRef]

32. Li, Y.; Wang, Z.; Xia, T.; Ju, H.; Zhang, K.; Long, R.; Xu, Q.; Wang, C.; Song, L.; Zhu, J.; et al. Implementing Metal-to-Ligand Charge Transfer in Organic Semiconductor for Improved Visible-near-Infrared Photocatalysis. *Adv. Mater.* **2016**, *28*, 6959–6965. [CrossRef]

33. Thomas, A.; Fischer, A.; Goettmann, F.; Antonietti, M.; Müller, J.-O.; Schlögl, R.; Carlsson, J.M. Graphitic Carbon Nitride Materials: Variation of Structure and Morphology and Their Use as Metal-Free Catalysts. *J. Mater. Chem.* **2008**, *18*, 4893–4908. [CrossRef]

34. Vorobyeva, E.; Chen, Z.; Mitchell, S.; Leary, R.K.; Midgley, P.; Thomas, J.M.; Hauert, R.; Fako, E.; Lopez, N.; Perez-Ramirez, J. Tailoring the Framework Composition of Carbon Nitride to Improve the Catalytic Efficiency of the Stabilised Palladium Atoms. *J. Mater. Chem. A* **2017**, *5*, 16393–16403. [CrossRef]

35. Chesnokov, V.V.; Kriventsov, V.V.; Malykhin, S.E.; Svintsitskiy, D.A.; Podyacheva, O.Y.; Lisitsyn, A.S.; Richards, R.M. Nature of Active Palladium Sites on Nitrogen Doped Carbon Nanofibers in Selective Hydrogenation of Acetylene. *Diam. Relat. Mater.* **2018**, *89*, 67–73. [CrossRef]

36. Arrigo, R.; Schuster, M.E.; Xie, Z.; Yi, Y.; Wowsnick, G.; Sun, L.L.; Hermann, K.E.; Friedrich, M.; Kast, P.; Hävecker, M.; et al. Nature of the N–Pd Interaction in Nitrogen-Doped Carbon Nanotube Catalysts. *ACS Catal.* **2015**, *5*, 2740–2753. [CrossRef]

37. Stonkus, O.A.; Kibis, L.S.; Yu. Podyacheva, O.; Slavinskaya, E.M.; Zaikovskii, V.I.; Hassan, A.H.; Hampel, S.; Leonhardt, A.; Ismagilov, Z.R.; Noskov, A.S.; et al. Palladium Nanoparticles Supported on Nitrogen-Doped Carbon Nanofibers: Synthesis, Microstructure, Catalytic Properties, and Self-Sustained Oscillation Phenomena in Carbon Monoxide Oxidation. *ChemCatChem* **2014**, *6*, 2115–2128. [CrossRef]

38. Brun, M.; Berthet, A.; Bertolini, J.C. XPS, AES and Auger Parameter of Pd and PdO. *J. Electron Spectrosc. Relat. Phenom.* **1999**, *104*, 55–60. [CrossRef]

39. Kibis, L.S.; Titkov, A.I.; Stadnichenko, A.I.; Koscheev, S.V.; Boronin, A.I. X-ray Photoelectron Spectroscopy Study of Pd Oxidation by RF Discharge in Oxygen. *Appl. Surf. Sci.* **2009**, *255*, 9248–9254. [CrossRef]

Article

Effects of the Carbon Support Doping with Nitrogen for the Hydrogen Production from Formic Acid over Ni Catalysts

Alina D. Nishchakova [1], Dmitri A. Bulushev [1,2,3], Olga A. Stonkus [2,3], Igor P. Asanov [1,3], Arcady V. Ishchenko [2,3], Alexander V. Okotrub [1,3] and Lyubov G. Bulusheva [1,3,*]

[1]   Laboratory of Physics & Chemistry of Nanomaterials, Nikolaev Institute of Inorganic Chemistry, SB RAS, 630090 Novosibirsk, Russia; nishchakova@niic.nsc.ru (A.D.N.); dmitri.bulushev@catalysis.ru (D.A.B.); asan@niic.nsc.ru (I.P.A.); spectrum@niic.nsc.ru (A.V.O.)

[2]   Boreskov Institute of Catalysis, SB RAS, 630090 Novosibirsk, Russia; stonkus@catalysis.ru (O.A.S.); arkady.ishchenko@gmail.com (A.V.I.)

[3]   Department of Natural Sciences, Novosibirsk State University, 630090 Novosibirsk, Russia

*   Correspondence: bul@niic.nsc.ru; Tel.: +7-383-330-5352

Received: 30 September 2019; Accepted: 23 October 2019; Published: 28 October 2019

Abstract: Porous nitrogen-doped and nitrogen-free carbon materials possessing high specific surface areas (400–1000 m^2 g^{-1}) were used for deposition of Ni by impregnation with nickel acetate followed by reduction. The nitrogen-doped materials synthesized by decomposition of acetonitrile at 973, 1073, and 1173 K did not differ much in the total content of incorporated nitrogen (4–5 at%), but differed in the ratio of the chemical forms of nitrogen. An X-ray photoelectron spectroscopy study showed that the rise in the synthesis temperature led to a strong growth of the content of graphitic nitrogen on the support accompanied by a reduction of the content of pyrrolic nitrogen. The content of pyridinic nitrogen did not change significantly. The prepared nickel catalysts supported on nitrogen-doped carbons showed by a factor of up to two higher conversion of formic acid as compared to that of the nickel catalyst supported on the nitrogen-free carbon. This was related to stabilization of Ni in the state of single Ni^{2+} cations or a few atoms clusters by the pyridinic nitrogen sites. The nitrogen-doped nickel catalysts possessed a high stability in the reaction at least within 5 h and a high selectivity to hydrogen (97%).

Keywords: nickel catalyst; porous carbon support; nitrogen doping; formic acid decomposition; hydrogen production

---

1. Introduction

Currently, hydrogen is one of the most promising substances that can replace usable sources of energy like natural gas or oil because it is ecologically pure and can be produced from renewable sources. At the same time, hydrogen usage has difficulties associated with its storage and transportation. That is why hydrogen production from different organic compounds with a relatively high content of hydrogen is a current trend in catalysis. Formic acid is a liquid organic hydrogen carrier containing 4.4 wt% of hydrogen, which can be liberated relatively easy using some catalysts. It is important that formic acid can be produced from biomass or CO_2 [1,2].

The most efficient catalysts for the hydrogen production from formic acid are based on palladium. However, noble metals are expensive and this is a reason for searching cheaper catalysts. Such catalysts could be based on Mo sulfide [3,4] or carbide [5], copper [6,7], or nickel [8–10].

Catalysts' supports may play an important role for catalytic activity especially when a highly dispersed active metal is used. Porous carbon materials are suitable candidates, since they have all the qualities that a support should possess. These materials demonstrate chemical, thermal, and mechanical stability and have a developed surface, which can be modified during the synthesis or post-synthetic processing by doping with heteroatoms like nitrogen [11,12]. They are usually cheap to synthesize and they are easy to handle. Our earlier results demonstrated a significant promotion of Pd catalysts by nitrogen doping of the carbon support for the gas-phase formic acid decomposition reaction [13]. This was related to the formation of new active sites—single Pd cations attached to pyridinic nitrogen species of the support. For the liquid phase decomposition, the nitrogen-doping played also a positive role [14,15].

Nickel is able to form single Ni atoms (cations) on the surface of nitrogen-doped carbon [16], which could be the active sites for some hydrogenation [17] and CO_2 electroreduction [18–21] reactions. It is not known whether these Ni species can be active in the formic acid decomposition reaction. Only theoretical grounds exist that single Ni cations inserted into carbon-nitrogen layer of the C_2N stoichiometry could be active in this reaction [22] providing energy barriers for the reaction steps comparable with those for Pt or Pd catalysts. Hence, the objective of our research was development of porous nitrogen-doped carbon supports for Ni catalysts of the hydrogen production from formic acid. Therefore, the decomposition of formic acid on four catalytic systems consisting of Ni deposited on the surfaces of porous nitrogen-doped and nitrogen-free carbon materials was studied.

2. Materials and Methods

2.1. Materials Synthesis

Porous carbon supports were synthesized using the chemical vapor deposition (CVD) method that is described in detail elsewhere [23]. The horizontal tubular quartz reactor was pumped out, filled with argon, and heated to the required temperature. Calcium tartrate doped with Fe (0.5 at%) was placed in the hot zone of the reactor for 420 s to produce template particles. After that, vapors of acetonitrile, taken as the source of nitrogen-doped carbon materials, or ethanol, used for the nitrogen-free material synthesis, were fed for 0.5 h. The product was kept in an HCl solution for 24 h, washed with distilled water on a polypropylene filter to pH = 7, and dried in an oven for an hour at 373 K. The synthesis was carried out at temperatures 973, 1073, and 1173 K and the obtained samples are referred to CN-973, CN-1073, CN-1173, and C-1073, where "CN" and "C" indicate the presence/absence of nitrogen atoms in the structure and "973", "1073", and "1173" is the synthesis temperature.

Heterogeneous nickel catalysts were synthesized by impregnation method: nickel precursor—$Ni(OAc)_2 \times 4H_2O$ (8.6 mg) was dissolved in tetrahydrofuran (6 mL) by stirring for 1200 s at room temperature. Then, the support (0.2 g) was added and the mixture was stirred for 4 h at 333 K. After that, tetrahydrofuran was removed by evaporation at room temperature. To decompose nickel acetate, the obtained pre-catalyst was heated at 623 K in an argon flow for 0.5 h and then cooled to room temperature without air access. All the weights were chosen to correspond to 1 wt% of deposited nickel.

2.2. Instrumental Methods

X-ray photoelectron spectroscopy (XPS) experiments were performed on a laboratory PHOIBOS 150 (SPECS GmbH, Berlin, Germany) spectrometer. Monochromatic radiation of the AlKα line (1486.7 eV) was used as an excitation source. An analysis of the porous structure was performed using an Autosorb iQ (Quantachrome Instruments, Boynton Beach, FL, USA) device at 77 K. A specimen was first treated in dynamic vacuum using standard 'outgas' option of the equipment at 373 K during 6 h. N_2 adsorption−desorption isotherms were measured within the range of relative pressures of 10^{-6} to 0.99. Then, specific surface areas were calculated using the Brunauer–Emmett–Teller (BET) method. The transmission electron microscopy (TEM) data were obtained using JEM-2010 and JEM-2200FS (JEOL Ltd., Tokyo, Japan) microscopes operated at 200 kV. Images with a high atomic number contrast were acquired using a high angle annular dark field (HAADF) detector in the Scanning-TEM (STEM) mode. However, this approach did not allow seeing single Ni atoms. The samples for the study were dispersed in ethanol by ultrasound and deposited on a holey carbon film mounted on a copper grid.

2.3. Catalytic Measurements

Formic acid decomposition was performed in a fixed-bed glass reactor, located in the furnace, using a flow set-up described in [24]. Argon was saturated with formic acid vapors passing through a glass container (bubbler) with liquid formic acid. After an additional dilution with argon, the concentration of formic acid vapor reached 2.5 vol% and the total gas flow rate was equal to 4020 mL s^{-1}. A catalyst was put into the reactor over a piece of quartz wool. Before every catalytic reaction experiment, the catalyst was heated in the same formic acid/Ar flow at 623 K for 0.5 h to reduce Ni and stabilize the catalyst.

Products of the decomposition reaction were analyzed by a Chromos GC-1000 gas chromatograph (Chromos Engineering, Moscow, Russia). The conversion of formic acid (X) was determined as the ratio of the sum of the obtained concentrations of CO and H_2 to the initial concentration of formic acid:

$$X = \frac{C_{CO} + C_{H_2}}{C_{HCOOH}} \times 100\%. \tag{1}$$

From the obtained conversion values, the specific reaction rates (W) were calculated, related to the mass of Ni in the catalyst:

$$W = \frac{X \times C_{HCOOH} \times V_f \times N_A}{22,400 \times m_{Ni}} \quad [\text{molecule s}^{-1}\text{g}_{Ni}^{-1}], \tag{2}$$

where C_{HCOOH}—initial formic acid concentration in the Ar flow, V_f—total flow rate (mL s^{-1}), N_A—the Avogadro constant (is equal to 6.022×10^{23} molecule mol^{-1}), m_{Ni}—Ni mass in the catalyst sample (g), and 22,400—volume of one mole of gas at 273 K and 1 atm (mL mol^{-1}).

3. Results and Discussion

3.1. Characterization of the Supports

A quick thermolysis of calcium tartrate accompanying by release of CO_2 and H_2O gases results in the formation of highly dispersed CaO particles [23]. These particles serve as a template for the growth of graphene-like layers catalyzed by iron incorporated into the template. Removal of CaO by washing of the synthesis product in HCl produces pores of different sizes. BET surface area values changed from 966 m^2 g^{-1} for the C-1073 sample to 407 m^2 g^{-1} for the CN-1173 sample (Table 1). All the nitrogen-doped carbon supports synthesized from acetonitrile had a lower BET surface area as compared to the nitrogen-free material synthesized from ethanol.

Table 1. Some characteristics of the supports and catalysts.

Support	BET Surface Area, m^2 g^{-1}	Content of N, at% (XPS)	Content of O, at% (XPS)	Catalyst	Mean Ni Particle Size, nm
C-1073	966	–	3.0 ± 0.5	Ni/C-1073	3.9 ± 1.2
CN-973	866	5.2 ± 0.5	6.2 ± 0.5	Ni/CN-973	<3 nm (rarely seen)
CN-1073	443	4.4 ± 0.5	3.2 ± 0.5	Ni/CN-1073	<3 nm (rarely seen)
CN-1173	407	4.6 ± 0.5	2.2 ± 0.5	Ni/CN-1173	5.5 ± 1.5

The nitrogen and oxygen contents in the supports were determined from the ratio of the areas under the C 1s, N 1s, and O 1s peaks in survey XPS data taking into account the photoionization cross-sections for the elements at the given photon energy. The obtained values are collected in Table 1. The content of nitrogen in the nitrogen-containing samples remains almost unchanged, corresponding to ca. 5 at%. The oxygen content in the samples decreases with the synthesis temperature from ca. 6.2 at% in CN-973 to ca. 2.2 at% in CN-1173.

The forms of the incorporated nitrogen were revealed from fitting of the N 1s spectra (Figure 1a). Four components were assigned to pyridinic (398.3 eV), pyrrolic (400.0 eV), graphitic (401 eV), and oxidized (402.5 eV) nitrogen [25]. Graphitic nitrogen atoms replaced the three-coordinated carbon atoms of graphene lattice, while other nitrogen species were formed at graphene edges and vacancies. Relative content of the nitrogen forms in the CN-supports is presented in Figure 1b. The rise of the synthesis temperature caused a decrease of the contents of pyrrolic nitrogen and an increase of the content of graphitic nitrogen, which was the most thermodynamically stable nitrogen form. The content of pyridinic nitrogen did not change significantly with the synthesis temperature.

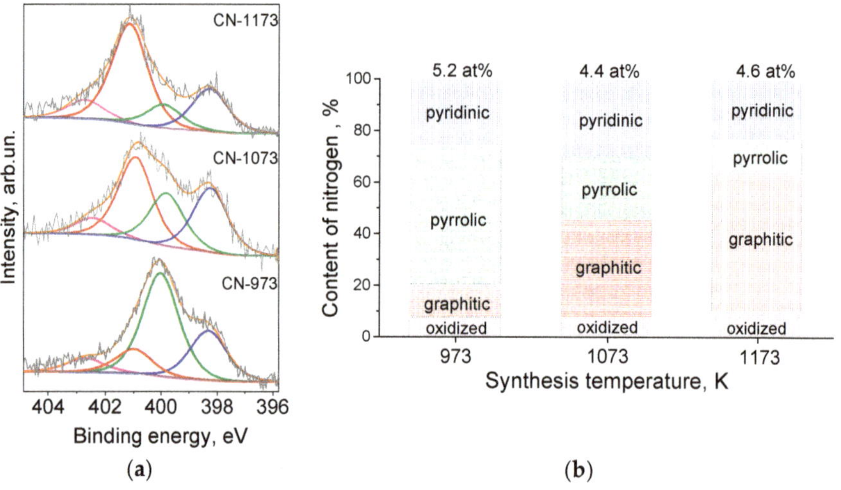

Figure 1. (a) N 1s X-ray photoelectron spectroscopy lines for the catalysts supports and (b) dependences of the content of nitrogen forms on the synthesis temperature.

3.2. Characterization of the Catalysts

Figure 2 shows an HAADF/STEM image of the Ni/C-1073 catalyst after the reaction. The mean particle size in this catalyst determined from these data was 3.9 nm (Table 1). Figure 3 shows TEM images of the Ni catalysts on the nitrogen-doped supports after the reaction. The only catalyst in which Ni containing particles were well seen was Ni/CN-1173. The mean particle size in this catalyst was 5.5 nm. In the Ni/CN-973 and Ni/CN-1073 catalysts, nanoparticles were rare and cannot be clearly seen in TEM images (Figure 3). The HAADF/STEM measurements of the Ni/CN-1073 sample found many extended regions in the sample without nanoparticles like that is shown in Figure 4a. However, the particles with a size smaller than 3 nm could be still found in this sample (Figure 4b,c), but this was a rare case. The smallest nickel species detected by HAADF/STEM had a size of 0.3–0.5 nm (Figure 4c). Hence, it was not possible to plot particle size distributions for these samples. The main part of Ni must be present in a highly dispersed state either in the form of single Ni atoms or few-atom clusters, which were almost not possible to observe by the techniques used. The difference between Ni/CN-1073 and Ni/C-1073 samples could be explained by a strong interaction of single Ni atoms with nitrogen species of the nitrogen-doped supports. In accordance, Zhong et al. [22] using density functional theory (DFT) calculations showed a strong interaction of a Ni atom with two pyridinic atoms of a C_2N fragment limiting Ni diffusion between different sites of the support and, therefore, preventing Ni sintering.

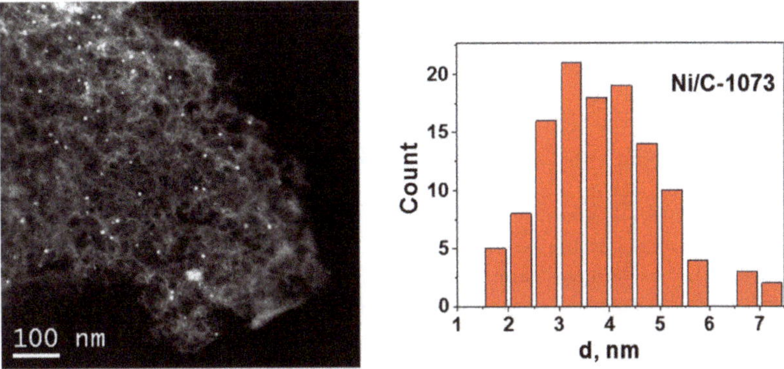

Figure 2. High angle annular dark field/scanning TEM (HAADF/STEM) image of the Ni/C-1073 catalyst after the reaction and Ni particle size distribution for this sample.

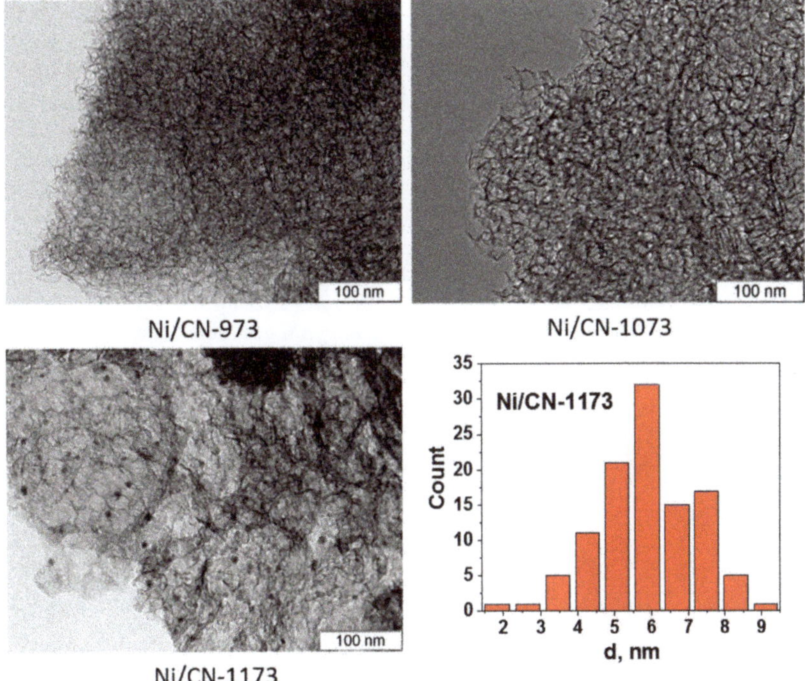

Figure 3. TEM images of the Ni catalysts, which were produced using nitrogen-doped porous carbon materials synthesized at 973 K (Ni/CN-973), 1073 K (Ni/CN-1073) and 1173 K (Ni/CN-1173), after the reaction and Ni particle size distribution for Ni/CN-1173.

Interesting that the increase of the synthesis temperature of the nitrogen-doped carbon up to 1173 K leads to an increase of the mean Ni particle size on the support. It becomes even bigger than that for the Ni/C-1073 catalyst. This could be related to the observed decrease of the content of oxygen and pyrrolic nitrogen species (Table 1). Besides, the TEM data showed that the Ni/CN-1173 sample has a smoother surface than the other two nitrogen-doped samples. The better graphitization of the support in the former case was due to the higher temperature of the material synthesis. This could be also a reason of the bigger size of Ni nanoparticles in the Ni/CN-1173 sample as compared to those in the Ni/CN-973 and Ni/CN-1073 samples. The observed change in the BET surface area should be also taken into account for explanation of the difference in the Ni dispersion for the Ni/C-1073 and Ni/CN-1173 samples.

The Ni $2p_{3/2}$ lines obtained by XPS for all catalysts are compared in Figure 5. It is seen that the binding energies of the Ni $2p_{3/2}$ lines differ for the nitrogen-doped and nitrogen-free catalysts. Thus, the position for the nitrogen-free catalyst corresponds to 856.3 eV while those for the nitrogen-doped catalysts synthesized at 973 and 1073 K are by about 1 eV lower (855.3 eV). Both positions should be assigned to oxidized Ni state (Ni^{2+}). Yamada et al. [16] assigned nickel with the Ni $2p_{3/2}$ binding energy at 855.5 eV to Ni cations interacting with pyridinic nitrogen atoms of the nitrogen-doped carbon support while that at 856.2 eV to $Ni(OH)_2$ and Ni interacting with oxygen containing functional groups, which are also presented in our nitrogen-free carbon support (Table 1). Similarly, Liu et al. [17] assigned their 855.3 eV line to single Ni cations attached to pyridinic nitrogen species of the nitrogen-doped carbon support since the catalyst did not contain nickel in the form of nanoparticles, but contained nickel only in the state of single atoms according to extended X-ray absorption fine structure (EXAFS) and HAADF/STEM data. This binding energy is also typical for different nickel(II) porphyrin derivatives [26]. The line at 855.3–855.5 eV was the most clearly seen for the Ni/CN-1073 sample implying that single Ni^{2+} cations attached to pyridinic nitrogen were the main species representing Ni in this catalyst. The line at 856.3 eV for the nitrogen-free Ni/C-1073 catalyst should be assigned mainly to $Ni(OH)_2$ in the nanoparticles observed by HAADF/STEM (Figure 2). The intermediate binding energy for the Ni/CN-1173 catalyst was explained by the presence of both Ni^{2+} species.

Hence, the electronic state of Ni in the nitrogen-doped catalysts differed from that in the nitrogen-free catalyst indicating interaction of Ni with nitrogen sites of the support. The nitrogen-doped catalysts mainly contained single Ni^{2+} cations attached to pyridinic nitrogen species of the support in agreement with the literature data.

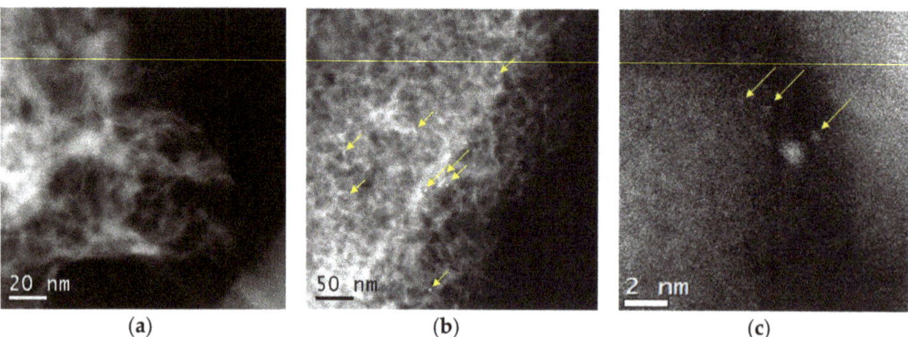
(a)   (b)   (c)

Figure 4. HAADF/STEM images of the Ni/CN-1073 catalyst after the reaction: (**a**) region without nanoparticles, (**b**) and (**c**) regions with nanoparticles and clusters.

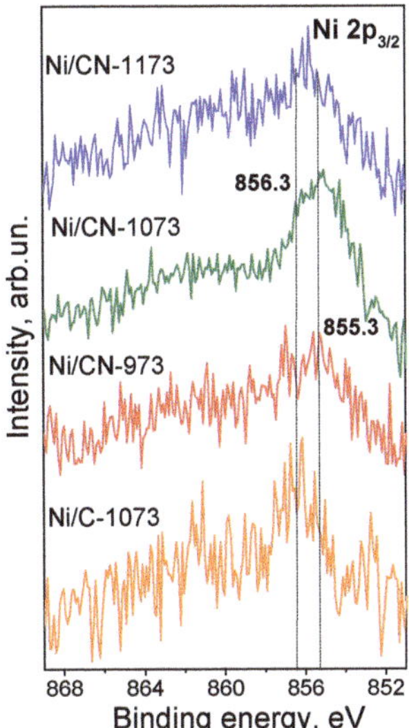

Figure 5. Ni 2p$_{3/2}$ X-ray photoelectron spectra of the Ni catalysts after the reaction.

3.3. Catalytic Studies

The conversion-temperature dependences are shown in Figure 6a. At low temperatures, the Ni/CN-973, Ni/CN-1073, and Ni/CN-1173 catalysts show a higher conversion than the Ni/C-1073 catalyst. Hence, the nitrogen insertion into the support promotes the reaction. The specific reaction rates calculated at 553 K are given in Figure 6b. This graph additionally shows the effect of the synthesis temperature on the catalytic activity. Thus, the temperature of 1073 K was optimal for the catalyst synthesis from acetonitrile. Since the activity and content of pyridinic nitrogen did not differ much, the growth of the activity with nitrogen should be assigned to the appearance of nickel sites stabilized by pyridinic nitrogen sites. The obtained activity was difficult to compare with the activity reported in the literature for some supported Ni catalysts [10], since the content of Ni in our samples was by a factor of 2–22 lower and the reaction temperature was higher (533–593 K vs. 400–500 K).

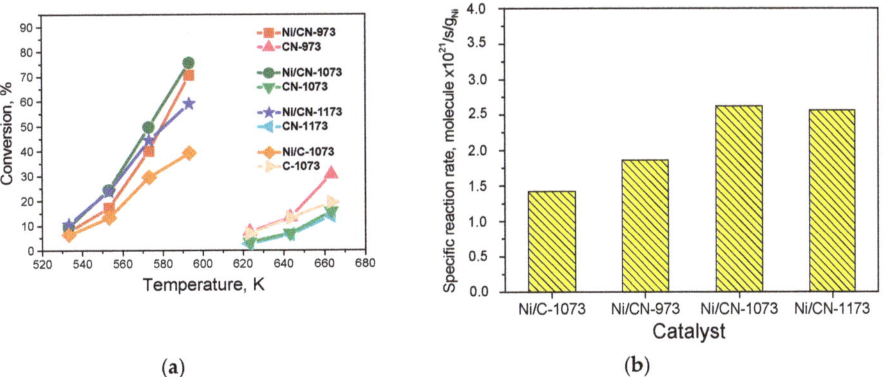

(**a**)   (**b**)

Figure 6. (**a**) Dependences of the formic acid decomposition conversions on temperature for the Ni catalysts (7 mg) and carbon supports (4 mg) and (**b**) dependence of the specific reaction rates for the Ni catalysts at 553 K.

The supports also could work as catalysts in the formic acid decomposition reaction, but not as efficiently, as Ni catalysts did (Figure 6a). At 663 K, the highest conversion was demonstrated by the CN-973 support (30%) and the lowest—by the CN-1173 support (14%). Such a spread in the conversions may be due to the differences in the specific surface areas of the supports (Table 1) and the presence of different nitrogen sites (Figure 1).

It is interesting that the activities of the Ni/CN-1073 and Ni/CN-1173 catalysts were close despite the mean Ni particle sizes were different (Table 1). It should be emphasized that used supports differed in the content of oxygen (Table 1) and in the ratio of graphitic and pyrrolic nitrogen species (Figure 1b). However, for better explanation of this result a determination of the ratio of atomic Ni species and Ni atoms present in nanoparticles is needed. DFT calculations of the formic acid interaction with different Ni structures would help also.

The apparent activation energies for the reaction over the studied catalysts were estimated from the Arrhenius plots. The obtained values were in the range of 100 ± 9 kJ mol^{-1}. Interesting that the close values were obtained earlier for formic acid decomposition over unsupported Ni powder and Ni catalysts supported on silica and alumina [10] as well as for decomposition of Ni formate [27]. The determined apparent activation energies for formic acid decomposition over carbon supports without Ni were significantly higher (125–148 kJ mol^{-1}).

To test the stability, the most promising catalyst Ni/CN-1073 was used. The results of the test for 5 h at 573 K are shown in Figure 7. It is seen that the hydrogen production selectivity was about 97%. The conversion demonstrates a slight increase from 57% to 61% in the time range from 20 to 60 min, but it did not change during the next 4 h. Thus, it was possible to conclude that this catalyst was quite stable and selective in the reaction, which means that there was no need to perform recycling experiments.

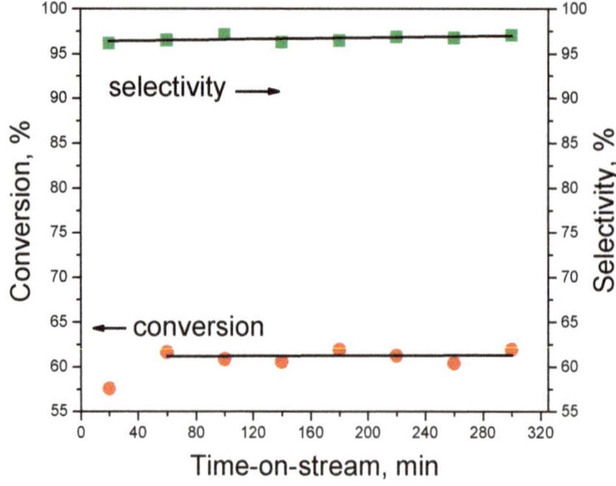

Figure 7. Stability test with the Ni/CN-1073 catalyst at 573 K.

4. Conclusions

Nitrogen-doping of the carbon support led to a significant increase of the rate of the hydrogen production from formic acid decomposition (by up to a factor of 2). The nitrogen-doped Ni catalysts showed very stable activity and high selectivity of the hydrogen production (97%). For the Ni catalysts on the nitrogen-doped supports synthesized at 973 K and 1073 K, it was difficult to find Ni containing nanoparticles by HAADF/STEM and TEM since the Ni dispersion was very high. Nickel in these catalysts was probably presented in the form of single Ni cations or few atoms clusters attached to pyridinic nitrogen of the support as was confirmed by XPS study. At the same time, the nanoparticles with a mean size of 3.9 nm were clearly observed in the Ni/C-1073 catalyst. The used nitrogen-doped supports had no significant difference in nitrogen content. The difference was in the content of each nitrogen form. This ratio of nitrogen forms and surface area of the support determine the Ni particle size. These parameters are interconnected and for better understanding of the role of particular parameter, it is necessary to perform EXAFS studies of the Ni state in the catalysts and DFT calculations of the possible Ni structures on nitrogen-doped carbon and interaction of such structures with the formic acid molecule.

Author Contributions: Synthesis, measurements and writing, A.D.N.; characterization, O.A.S., A.V.I. and I.P.A.; funding acquisition, supervision and data curation, L.G.B. and A.V.O.; methodology, data analysis and writing, D.A.B.

Funding: This research was funded by the Russian Science Foundation, grant number 16-13-00016.

Acknowledgments: The XPS studies were conducted using the equipment of the Center of Collective Use «National Center of Catalyst Research».

Conflicts of Interest: The authors declare no conflict of interest.

Nomenclature

BET	Brunauer–Emmett–Teller
DFT	Density Functional Theory
EXAFS	Extended X-ray Absorption Fine Structure
HAADF/STEM	high angle annular dark field/scanning transmission electron microscopy
HRTEM	high resolution transmission electron microscopy
XPS	X-ray photoelectron spectroscopy

References

1. Bulushev, D.A.; Ross, J.R.H. Towards Sustainable Production of Formic Acid. *ChemSusChem* **2018**, *11*, 821–836. [CrossRef] [PubMed]
2. Bulushev, D.A.; Ross, J.R.H. Heterogeneous Catalysts for Hydrogenation of CO_2 and Bicarbonates to Formic Acid and Formates. *Catal. Rev.* **2018**, *60*, 566–593. [CrossRef]
3. Koroteev, V.O.; Bulushev, D.A.; Chuvilin, A.L.; Okotrub, A.V.; Bulusheva, L.G. Nanometer-Sized MoS2 Clusters on Graphene Flakes for Catalytic Formic Acid Decomposition. *ACS Catal.* **2014**, *4*, 3950–3956. [CrossRef]
4. Kurnia, I.; Yoshida, A.; Situmorang, Y.A.; Kasai, Y.; Abudula, A.; Guan, G. Utilization of Dealkaline Lignin as a Source of Sodium-Promoted MoS_2/Mo_2C Hybrid Catalysts for Hydrogen Production from Formic Acid. *ACS Sustain. Chem. Eng.* **2019**, *7*, 8670–8677. [CrossRef]
5. Koós, Á.; Solymosi, F. Production of CO-Free H_2 by Formic Acid Decomposition over Mo_2C/Carbon Catalysts. *Catal. Lett.* **2010**, *138*, 23–27. [CrossRef]
6. Bulushev, D.A.; Chuvilin, A.L.; Sobolev, V.I.; Stolyarova, S.G.; Shubin, Y.V.; Asanov, I.P.; Ishchenko, A.V.; Magnani, G.; Riccò, M.; Okotrub, A.V.; et al. Copper on Carbon Materials: Stabilization by Nitrogen Doping. *J. Mater. Chem. A* **2017**, *5*, 10574–10583. [CrossRef]
7. Pechenkin, A.; Badmaev, S.; Belyaev, V.; Sobyanin, V. Production of Hydrogen-Rich Gas by Formic Acid Decomposition over CuO-CeO_2/γ-Al_2o_3 Catalyst. *Energies* **2019**, *12*, 3577. [CrossRef]
8. Bing, Q.M.; Liu, W.; Yi, W.C.; Liu, J.Y. Ni Anchored C2N Monolayers as Low-Cost and Efficient Catalysts for Hydrogen Production from Formic Acid. *J. Power Sources* **2019**, *413*, 399–407. [CrossRef]
9. Fujitsuka, H.; Nakagawa, K.; Hanprerakriengkrai, S.; Nakagawa, H.; Tago, T. Hydrogen Production from Formic Acid Using Pd/C, Pt/C, and Ni/C Catalysts Prepared from Ion-Exchange Resins. *J. Chem. Eng. Jpn.* **2019**, *52*, 423–429. [CrossRef]
10. Iglesia, E.; Boudart, M. Decomposition of Formic Acid on Copper, Nickel, and Copper-Nickel Alloys: Iii. Catalytic Decomposition on Nickel and Copper-Nickel Alloys. *J. Catal.* **1983**, *81*, 224–238. [CrossRef]
11. Yang, G.; Han, H.; Li, T.; Du, C. Synthesis of Nitrogen-Doped Porous Graphitic Carbons Using Nano-$CaCO_3$ as Template, Graphitization Catalyst, and Activating Agent. *Carbon* **2012**, *50*, 3753–3765. [CrossRef]

12. Xia, Y.; Mokaya, R. Synthesis of Ordered Mesoporous Carbon and Nitrogen-Doped Carbon Materials with Graphitic Pore Walls Via a Simple Chemical Vapor Deposition Method. *Adv. Mater.* **2004**, *16*, 1553–1558. [CrossRef]

13. Zacharska, M.; Bulusheva, L.G.; Lisitsyn, A.S.; Beloshapkin, S.; Guo, Y.; Chuvilin, A.L.; Shlyakhova, E.V.; Podyacheva, O.Y.; Leahy, J.J.; Okotrub, A.V.; et al. Factors Influencing the Performance of Pd/C Catalysts in the Green Production of Hydrogen from Formic Acid. *ChemSusChem* **2017**, *10*, 720–730. [CrossRef] [PubMed]

14. Navlani-García, M.; Mori, K.; Salinas-Torres, D.; Kuwahara, Y.; Yamashita, H. New Approaches toward the Hydrogen Production from Formic Acid Dehydrogenation over Pd-Based Heterogeneous Catalysts. *Front. Mater.* **2019**, *6*. [CrossRef]

15. Salinas-Torres, D.; Navlani-García, M.; Mori, K.; Kuwahara, Y.; Yamashita, H. Nitrogen-Doped Carbon Materials as a Promising Platform toward the Efficient Catalysis for Hydrogen Generation. *Appl. Catal. A Gen.* **2018**, *571*, 25–41. [CrossRef]

16. Yamada, Y.; Suzuki, Y.; Yasuda, H.; Uchizawa, S.; Hirose-Takai, K.; Sato, Y.; Suenaga, K.; Sato, S. Functionalized Graphene Sheets Coordinating Metal Cations. *Carbon* **2014**, *75*, 81–94. [CrossRef]

17. Liu, W.; Chen, Y.; Qi, H.; Zhang, L.; Yan, W.; Liu, X.; Yang, X.; Miao, S.; Wang, W.; Liu, C.; et al. A Durable Nickel Single-Atom Catalyst for Hydrogenation Reactions and Cellulose Valorization under Harsh Conditions. *Angew. Chem. Int. Ed.* **2018**, *57*, 7071–7075. [CrossRef]

18. Jiang, K.; Siahrostami, S.; Akey, A.J.; Li, Y.; Lu, Z.; Lattimer, J.; Hu, Y.; Stokes, C.; Gangishetty, M.; Chen, G.; et al. Transition-Metal Single Atoms in a Graphene Shell as Active Centers for Highly Efficient Artificial Photosynthesis. *Chem* **2017**, *3*, 950–960. [CrossRef]

19. Cheng, Y.; Zhao, S.; Johannessen, B.; Veder, J.-P.; Saunders, M.; Rowles, M.R.; Cheng, M.; Liu, C.; Chisholm, M.F.; Marco, R.; et al. Atomically Dispersed Transition Metals on Carbon Nanotubes with Ultrahigh Loading for Selective Electrochemical Carbon Dioxide Reduction. *Adv. Mater.* **2018**, *30*, 1706287. [CrossRef]

20. Yang, J.; Qiu, Z.; Zhao, C.; Wei, W.; Chen, W.; Li, Z.; Qu, Y.; Dong, J.; Luo, J.; Li, Z.; et al. In Situ Thermal Atomization to Convert Supported Nickel Nanoparticles into Surface-Bound Nickel Single-Atom Catalysts. *Angew. Chem.* **2018**, *130*, 14291–14296. [CrossRef]

21. Cheng, Y.; Zhao, S.; Li, H.; He, S.; Veder, J.-P.; Johannessen, B.; Xiao, J.; Lu, S.; Pan, J.; Chisholm, M.F.; et al. Unsaturated Edge-Anchored Ni Single Atoms on Porous Microwave Exfoliated Graphene Oxide for Electrochemical CO_2. *Appl. Catal. B Environ.* **2019**, *243*, 294–303. [CrossRef]

22. Zhong, W.; Liu, Y.; Deng, M.; Zhang, Y.; Jia, C.; Prezhdo, O.V.; Yuan, J.; Jiang, J. C2N-Supported Single Metal Ion Catalysts for HCOOH Dehydrogenation. *J. Mater. Chem. A* **2018**, *6*, 11105–11112. [CrossRef]

23. Shlyakhova, E.V.; Bulusheva, L.G.; Kanygin, M.A.; Plyusnin, P.E.; Kovalenko, K.A.; Senkovskiy, B.V.; Okotrub, A.V. Synthesis of Nitrogen-Containing Porous Carbon Using Calcium Oxide Nanoparticles. *Phys. Status Solidi B* **2014**, *251*, 2607–2612. [CrossRef]

24. Golub, F.S.; Beloshapkin, S.; Gusel'nikov, A.V.; Bolotov, V.A.; Parmon, V.N.; Bulushev, D.A. Boosting Hydrogen Production from Formic Acid over Pd Catalysts by Deposition of N-Containing Precursors on the Carbon Support. *Energies* **2019**, *12*, 3885. [CrossRef]

25. Okotrub, A.V.; Fedorovskaya, E.O.; Senkovskiy, B.V.; Bulusheva, L.G. Nitrogen Species in Few-Layer Graphene Produced by Thermal Exfoliation of Fluorinated Graphite Intercalation Compounds. *Phys. Status Solidi B* **2015**, *252*, 2444–2450. [CrossRef]

26. Berríos, C.; Cárdenas-Jirón, G.I.; Marco, J.F.; Gutiérrez, C.; Ureta-Zañartu, M.S. Theoretical and Spectroscopic Study of Nickel(Ii) Porphyrin Derivatives. *J. Phys. Chem. A* **2007**, *111*, 2706–2714. [CrossRef]

27. Iglesia, E.; Boudart, M. Decomposition of Formic-Acid on Copper, Nickel, and Copper Nickel-Alloys. IV. Temperature-Programmed Decomposition of Bulk Nickel Formate and of Formic-Acid Preadsorbed on Nickel Powder. *J. Catal.* **1984**, *88*, 325–332. [CrossRef]

Article

Nitrogen Doped Carbon Nanotubes and Nanofibers for Green Hydrogen Production: Similarities in the Nature of Nitrogen Species, Metal–Nitrogen Interaction, and Catalytic Properties

Olga Podyacheva [1,2,*], Alexander Lisitsyn [1], Lidiya Kibis [1,2], Andrei Boronin [1,2], Olga Stonkus [1,2], Vladimir Zaikovskii [1,2], Arina Suboch [1], Vladimir Sobolev [1] and Valentin Parmon [1,2]

[1] Boreskov Institute of Catalysis, SB RAS, Novosibirsk 630090, Russia; liss@catalysis.ru (A.L.);
    kibis@catalysis.ru (L.K.); boronin@catalysis.ru (A.B.); stonkus@catalysis.ru (O.S.); viz@catalysis.ru (V.Z.);
    arina@catalysis.ru (A.S.); visobo@catalysis.ru (V.S.); parmon@catalysis.ru (V.P.)
[2] Novosibirsk State University, Novosibirsk 630090, Russia
*   Correspondence: pod@catalysis.ru

Received: 20 September 2019; Accepted: 17 October 2019; Published: 18 October 2019

Abstract: The effect of nitrogen doped bamboo-like carbon nanotubes (N–CNTs) on the properties of supported platinum (0.2 and 1 wt %) catalysts in formic acid decomposition for hydrogen production was studied. It was shown that both impregnation and homogeneous precipitation routes led to the formation of electron-deficient platinum stabilized by pyridinic nitrogen sites of the N–CNTs. The electron-deficient platinum species strongly enhanced the activity and selectivity of the Pt/N–CNTs catalysts when compared to the catalysts containing mainly metallic platinum nanoparticles. A comparison of bamboo-like N–CNTs and herring-bone nitrogen doped carbon nanofibers (N–CNFs) as the catalyst support allowed us to conclude that the catalytic properties of supported platinum are determined by its locally one-type interaction with pyridinic nitrogen sites of the N–CNTs or N–CNFs irrespective of substantial structural differences between nanotubes and nanofibers.

Keywords: hydrogen; formic acid; platinum; nitrogen doped; carbon nanotubes; carbon nanofibers

---

1. Introduction

Hydrogen is considered as a promising energy source although its safe storage and transportation are rather complicated tasks. In this connection, the concept of hydrogen storage as a component of various hydrogen-rich compounds such as borohydrates, hydrazine, or formic acid has recently received considerable attention. Among such compounds, formic acid (FA) is of particular interest. First, HCOOH contains a large amount of hydrogen (4.4 wt %), and second, FA can be synthesized from renewable sources, for example, via different transformations of cellulose, which certainly makes HCOOH an attractive source of hydrogen [1].

It is known that the decomposition of FA on homogeneous or heterogeneous metal catalysts can proceed via two routes, leading to the formation of either H_2 and CO_2 or H_2O and CO. Thus, in the case of heterogeneous catalysts, the reaction rate and selectivity of hydrogen formation depend on the sizes, morphologies, distributions, and surface states of metal nanoparticles as well as the properties of catalyst support (CeO_2, SiO_2, C, C–N, etc.) [2]. Among the various types of supports for metal catalysts used for FA decomposition, carbon materials are of special interest because they allow the size and electronic state of supported metal nanoparticles to be controlled more precisely [3,4]. Moreover, the application of nitrogen doped carbon nanomaterials (N–CNMs) or carbon nitride makes it possible to obtain stable single atom species exhibiting high activity and selectivity in the FA decomposition reaction [5,6]. Stabilization of these highly active species proceeds via strong interaction of metal with pyridinic nitrogen centers of nitrogen doped carbon nanofibers (N–CNFs), nitrogen doped carbon nanotubes (N–CNTs), or C_3N_4, which has been confirmed by x-ray photoelectron spectroscopy (XPS) data and density functional theory (DFT) calculations [5,7–11].

According to the literature, the synthesis method substantially affects the particle size and electronic state of metals deposited on N–CNMs. For example, for the Pd catalyst supported on N–CNFs and nitrogen-doped porous carbon network, we demonstrated that Pd acetate is more preferable than Pd nitrate for the synthesis of a more dispersed, and therefore more active catalyst [7]. Moreover, when palladium was deposited from Pd acetate on N–CNTs in the amount of <0.8 wt %, 100% of palladium was represented by highly active isolated ions showing stability in both a hydrogen atmosphere at elevated temperatures and in the course of the HCOOH decomposition reaction [12]. The authors of [8] investigated the properties of 0.5–5% Pt/N–CNTs catalysts synthesized with H_2PtCl_6 as a precursor using different deposition methods: ethylene glycol reduction, sodium borohydride reduction, or impregnation-H_2-reduction techniques. It was found that the difference between synthesis routes was caused by the different metal–support interaction (MSI) between the Pt and nitrogen species, and the impregnation technique led to the strongest MSI via defects in N–CNTs comprising carbon vacancies and pyridinic nitrogen. As a result, a correlation between the binding energies of $Pt4f_{7/2}$ (E_b $Pt4f_{7/2}$) and activity in the electro-oxidation of glycerol and formic acid was found. Previously, we have shown that such types of defect can form in the graphene layer of N–CNTs [13].

Our earlier comparative study of Pt-group metals (Pt, Pd, Ru) deposited on herring-bone N–CNFs revealed that the activity of the catalysts toward formic acid decomposition decreases in the following order: Pt>Pd>Ru [5]. DFT calculations made it possible to propose the active site consisting of a metal atom coordinated by a pair of pyridinic nitrogen atoms at the edges of the graphene sheets. Indeed, it is known that multiple edges of graphene layers enriched with pyridinic nitrogen are segregated to the surface of N–CNFs [14], which explains the preferential interaction of deposited metals with these sites. In this connection, it seems interesting to explore the interaction of metals with N–CNMs with another structure, for example, with the widely used bamboo-like N–CNTs where all nitrogen species (pyridinic, pyrrolic, and graphitic) are located within internal and external graphene layers [15] and can equiprobably participate in the stabilization of deposited metals.

This paper is devoted to the investigation of the effect of the bamboo-like N–CNTs on the properties of Pt/N–CNTs catalysts synthesized by impregnation and homogeneous precipitation methods. The bamboo-like N–CNTs were found to affect the formation of electron-deficient platinum, showing a 3–4 fold higher activity in formic acid decomposition when compared to the metallic platinum nanoparticles. A comparative study of platinum deposited on bamboo-like N–CNTs and herring-bone N–CNFs revealed that pyridinic nitrogen plays a key role in enhancing the activity of the catalysts irrespective of the structural features of the carbon nanomaterial.

2. Materials and Methods

Nitrogen-free carbon nanotubes (CNTs) and bamboo-like nitrogen-doped N–CNTs were grown on an Fe–Ni-containing catalyst ($62\%Fe–8\%Ni–30\%Al_2O_3$) via decomposition of ethylene or 40% ethylene–60% ammonia mixture at 700 °C in a flow reactor with a fluidized catalyst bed [16]. Synthesized CNTs and N–CNTs were thoroughly washed to remove the growth catalyst first in concentrated HCl, then boiled in 2 M HCl for 6 h, and washed in distilled water until no chloride ions were detected in the rinsing liquid. The washed carbons were dried in Ar at 170 °C for 2 h.

Platinum (0.2 and 1 wt %) was deposited on the CNTs and N–CNTs by incipient wetness impregnation and homogeneous precipitation techniques using H_2PtCl_6 as the precursor. In the incipient wetness impregnation method, the defined volume of the H_2PtCl_6 aqueous solution was added to the freshly dried CNTs or N–CNTs at room temperature, the sample was thoroughly stirred, stored in a closed vessel for 30 min, and then dried in air for 24 h. The volume of the H_2PtCl_6 solution was defined based on the water capacity of the CNTs and N–CNTs [17]. The samples were designated as

0.2–2%Pt/CNTs-im and 0.2–2%Pt/N–CNTs-im. In the homogeneous precipitation technique, a known amount of H_2PtCl_6 was added to an aqueous suspension of the CNTs or N–CNTs and stirred for 60 min at room temperature. Then, a stoichiometric amount of 0.5 M solution of NaOH was added dropwise at room temperature. The mixture was stirred and heated for 1.5 h at 70 °C. Ultimately, the pH reached 8. The precipitates were filtered, carefully washed with distilled water, and dried in air. The samples were designated as 0.2–2%Pt/CNTs-pr and 0.2–2%Pt/N–CNTs-pr. Samples prepared by both methods were reduced in a 30 vol.% H_2/Ar mixture for 1 h at 250 °C. The content of Pt in the catalysts was determined by x-ray fluorescence spectroscopy.

Chemisorption measurements were made using a pulse technique and CO as the adsorbate at room temperature. Before the chemisorption measurements, the catalysts were re-reduced in situ in flowing hydrogen at 100 °C. The fractions of metal atoms exposed ($Pt_{surface}/Pt_{total}$) were calculated by assuming that the stoichiometry of adsorption $CO/Pt_{surface}$ was equal to 1.

High-angle annular dark-field (HAADF STEM) images were obtained on a JEM-2200FS electron microscope at an accelerating voltage of 200 kV using a high angle annular dark field (HAADF) detector in the scanning-TEM (STEM) mode.

X-ray photoelectron spectroscopy study was performed with an ES-300 (KRATOS Analytical) photoelectron spectrometer at a constant pass energy of a photoelectron energy analyzer. The spectra were recorded using an AlKα source with a quantum energy of 1486.6 eV. The energy scale was calibrated against the binding energy of gold $Au4f_{7/2}$ (equal to 84.0 eV) and copper $Cu2p_{3/2}$ (equal to 932.7 eV) [18]. The qualitative control of the surface chemical composition was made using the survey spectra in the range of 0–1100 eV with the analyzer pass energy HV = 50 eV and scan step 1 eV. To analyze the quantitative composition and chemical state of elements, core levels of elements were recorded (Pt4f, Al2p, Cl2p, C1s, O1s, N1s, and Fe2p) at the analyzer pass energy of HV = 25 eV and scan step of 0.1 eV. Since the Pt4f and Al2p lines overlap, control recording of the Pt4f + Al2p spectral region for the CNTs and N–CNTs supports was performed to take into account a possible contribution of the Al2p line of aluminum from the growth catalyst to the Pt4f spectrum of platinum. Prior to the experiments, the samples were treated ex situ with hydrogen at 250 °C for 1 h.

Experiments on formic acid decomposition (5 vol.% HCOOH/He) were performed in a flow set-up using a quartz reactor with the inner diameter of 6 mm. The catalyst loading was 20 mg; the catalyst powder was uniformly mixed with 0.5 cm^3 of quartz sand (fraction 0.25–0.5 mm). The reaction mixture feed rate was 20 cm^3/min. The experiments were carried out in the temperature-programmed mode, the temperature being raised at a rate of 2 deg/min, and with chromatographic analysis of the gas mixture. The temperature of the reactor was measured using a thermocouple placed in the catalyst bed. All gas lines were preheated in a special box to 100 °C, whereas the reactor placed in the furnace had a special zone where the incoming gas was preheated to the reaction temperature.

The turnover frequencies (TOF) of the 1%Pt/CNTs-pr and bamboo-like 1%Pt/N–CNTs-pr were compared with the TOFs of Pt catalysts deposited on the herring-bone carbon nanofibers (1%Pt/N–CNFs). The properties of the 1%Pt/N–CNFs catalysts were described in our previous article [19]. The TOF values for 1%Pt/N–CNFs catalysts for 125 °C were calculated using the experimental data presented in [19].

3. Results and Discussion

Temperature dependences of the FA conversion on carbon supports and on supported Pt catalysts synthesized by homogeneous precipitation route are displayed in Figure 1. One can see that CNTs and N–CNTs possess some intrinsic activity, which is a typical for carbon nanomaterials [20–24]. The deposition of 0.2 wt % platinum on both supports shifted the temperature curve of conversion toward low temperatures by more than 100 °C. Note that a greater shift was observed in the case of 0.2%Pt/N–CNTs, which indicates differences in the properties of deposited platinum when going from CNTs to N–CNTs. The enhancement of the activity of N-doped catalysts when compared to N-free catalysts became even more pronounced if 1 wt % of Pt was deposited on the carbon nanotubes, as seen in Figure 1b.

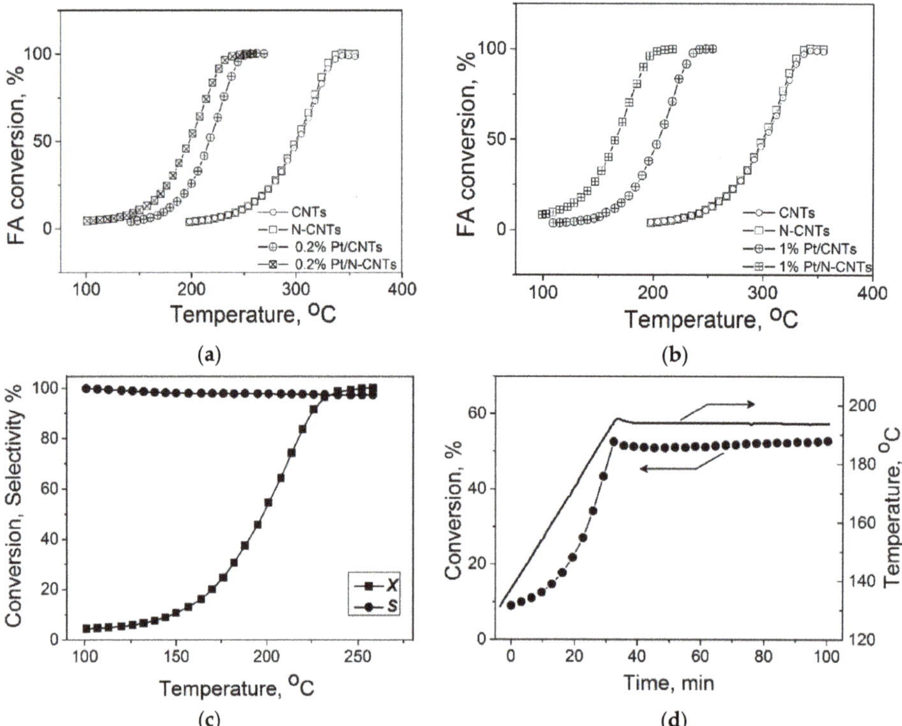

Figure 1. Temperature dependences of FA conversion on the CNTs, N–CNTs, and supported Pt catalysts synthesized by the homogeneous precipitation route: 0.2%Pt (**a**) and 1%Pt (**b**). Temperature dependences of FA conversion and selectivity on 0.2%Pt–N–CNTs-pr (**c**) and stability test of the 0.2%Pt–N–CNTs-im in the course of the reaction (**d**).

It should also be noted that the promoting effect of N–CNTs when compared to CNTs was observed irrespective of the deposition method (impregnation or homogeneous precipitation). In all cases, the TOF values for N-doped catalysts exceeded the TOFs for N-free catalysts by a factor of 3–4 (Figure 2). One can also see that the increase in TOF was accompanied by a substantial increase in the selectivity toward hydrogen formation, from 92.6 to 98%; this means a decrease of the CO content by ~4 times, where the latter is very important when FA is used for hydrogen production. Earlier, for the N–CNFs based catalysts, we demonstrated by DFT calculations that single metal ions coordinated by pyridinic nitrogen sites caused the cleavage of the C–H bond in the FA molecule to give adsorbed hydrogen atom and carboxyl fragments [5]. It was demonstrated that the adsorbed hydrogen atom and hydrogen of the carboxyl fragment were directed toward each other, thus providing an easy formation of hydrogen and CO_2.

At present, the improved properties of metal catalysts supported on various N–CNMs are attributed to a decrease in the particle size or an increase/decrease in the electron density of metals due to MSI via nitrogen species or defects consisting of carbon vacancies and nitrogen sites [25,26]. Indeed, since the specific areas of all of our catalysts measured by Brunauer-Emmett-Teller method varied in a very narrow range (140–150 m^2/g), the observed differences were most likely related to differences in the properties of the deposited platinum. To verify this hypothesis, various characteristics of deposited platinum were analyzed; first, sizes of the platinum particles in the supported catalysts were compared.

For this purpose, we used chemisorption of CO, which is sensitive to the dispersion of metals [27,28] as well as HAADF STEM.

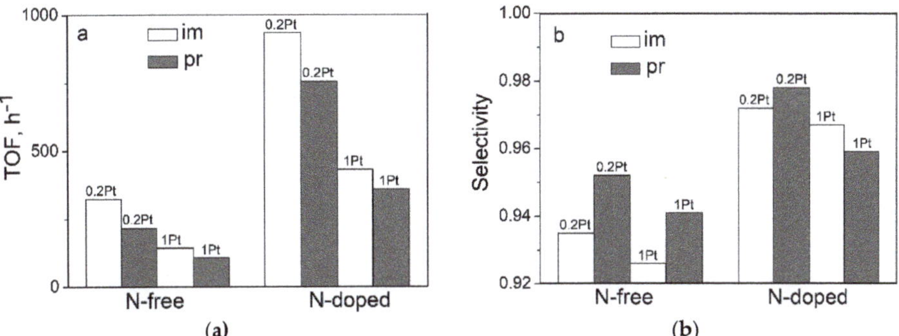

Figure 2. TOF values of the FA decomposition at 125 °C (**a**) and selectivity at 50% FA conversion (**b**) on the Pt catalysts deposited on CNTs and N–CNTs by different routes: im–impregnation, pr–homogeneous precipitation.

As seen in Figure 3, a minor increase in the TOF with an increase in the CO/Pt ratio was observed for the N-free catalysts, which may be assigned to similar dispersions of deposited metallic platinum in the tested catalysts [27,28]. In the case of N-doped catalysts, an inverse dependence was observed: TOF decreases with increases in the CO/Pt value. In addition, for this series of samples, a decrease in the platinum content from 1 to 0.2% was accompanied by a considerable (nearly 3-fold) decrease in the CO/Pt value, irrespective of the platinum deposition method. Such a pronounced decrease in the amount of chemisorbed CO by lowering the concentration of the deposited metal from 2 to 0.2 wt % was observed in our earlier study for the Pd/N–CNTs catalysts; a possible reason is that the catalysts with a palladium content <0.8 wt % completely consisted of isolated palladium ions, which poorly chemisorb CO [12].

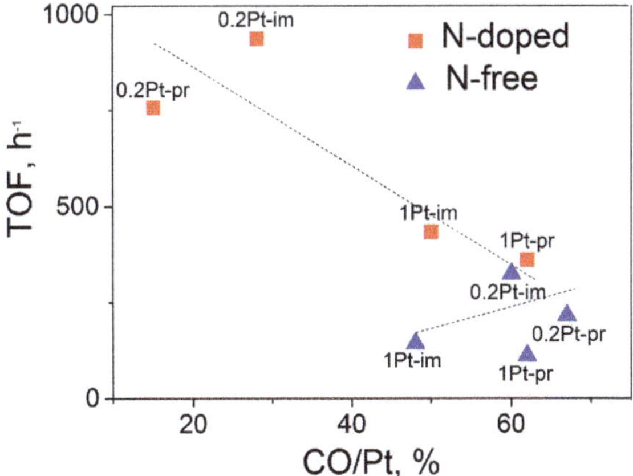

Figure 3. The influence of TOF of formic acid decomposition at 125 °C on the CO/Pt ratio for the N-doped and N-free catalysts differing in Pt content and preparation route: im–impregnation, pr–homogeneous precipitation.

To verify this assumption, the catalysts were studied by HAADF STEM. It was found that the most active catlyst 0.2%Pt/N–CNTs-im consisted of Pt particles with the size of about 1.1 nm and a narrow size distribution, whereas the lowest active 1%Pt/CNTs-pr contained nanoparticles of ~1.9 nm with a wider size distribution (Table 1). Meanwhile, the size of the nanoparticles in all samples, irrespective of the support type, deposition method, and Pt content, changed in a quite narrow range of 1.1–1.9 nm, which cannot explain such strong differences in the CO/Pt values from 15 to 67% (Figure 3). Therefore, it can be supposed that the catalysts contain platinum clusters or atoms/ions besides nanoparticles. Indeed, the presence of such species was discovered in all catalysts. The most representative photos of the catalysts showing single atoms are given in Figure 4. However, in the N-doped catalysts, the quantity of the isolated platinum atoms was found to be considerably higher, and these atoms were distributed on the N–CNT surface uniformly when compared with N-free catalysts.

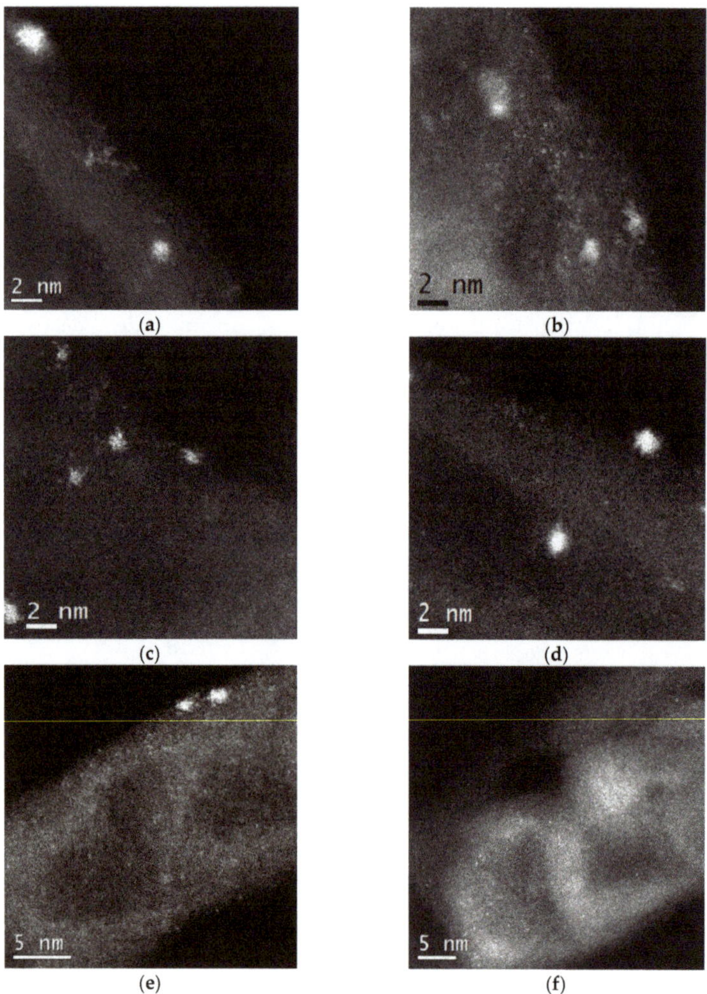

Figure 4. HAADF STEM images of 1%Pt/CNTs-imp (**a**), 1%Pt/N–CNTs-imp (**b**), 1%Pt/CNTs-pr (**c**), 1%Pt/N–CNTs-pr (**d**), 0.2%Pt/N–CNTs-imp (**e**), and 0.2%Pt/N–CNTs-pr (**f**).

Table 1. HAADF STEM platinum nanoparticle size (nm), standard deviation (sd), TOF (h^{-1}) at 125 °C.

Catalyst	Pt Size, nm	sd	TOF, h^{-1}
1%Pt/CNTs-im	1.3	0.4	144
1%Pt/CNTs-pr	1.9	0.7	108
0.2%Pt/CNTs-im	1.2	0.3	324
0.2%Pt/CNTs-pr	1.3	0.5	216
1%Pt/N–CNTs-im	1.4	0.4	432
1%Pt/N–CNTs-pr	1.3	0.3	360
0.2%Pt/N–CNTs-im	1.1	0.2	936
0.2%Pt/N–CNTs-pr	1.6	0.4	756

Thus, it can be concluded that both methods of platinum deposition on CNTs and N–CNTs can produce dispersed catalysts consisting of nanoparticles and isolated platinum atoms. For the N-free catalysts, the samples mainly consisted of nanoparticles and quite a low quantity of single atoms, which did not significantly influence the CO/Pt and TOF values (Figure 3). For the N-doped catalysts, irrespective of their preparation procedure, a decrease in the platinum content strongly lowered the CO/Pt ratio and increased the TOF in HCOOH decomposition and the selectivity of hydrogen formation.

Along with this, of note is a slightly higher activity of the catalysts obtained by incipient wetness impregnation when compared to the catalysts synthesized by homogeneous precipitation, irrespective of the support type and platinum content (Figure 2). The observed difference for the N-free catalysts can be attributed to a larger particle size and a broader particle size distribution of the catalysts obtained by homogeneous precipitation (Table 1). In the case of the impregnation method, platinum particles were formed directly from $[PtCl_6]^{2-}$ anions that were molecularly adsorbed on the carbon surface. In homogeneous precipitation from aqueous solutions, hydrolysis can lead to the formation of not only mono-, but also polynuclear platinum complexes $[Pt(OH)_xO_y]_n$ and hence to the formation of larger platinum particles with different sizes [27]. For the N-doped catalysts, the situation seems to be more complicated and in this case, the MSI between the Pt and nitrogen groups should similarly be taken into account like in [8]. However, the homogeneous precipitation method is preferable because it has no scaling limitations.

The difference between N-free and N-doped catalysts can also be seen in the XPS spectra. Figure 5a shows an example for 0.2%Pt/CNTs-pr and 0.2%Pt/N–CNTs-pr samples. In the case of 0.2%Pt/N–CNTs-pr, the Pt4f line shifted toward higher binding energies by ca. 0.3 eV. The difference spectrum shows a doublet with $E_b(Pt4f_{7/2}) = 72.9$ eV, which indicates the presence of two or more Pt states in the catalysts. Deconvolution of Pt4f spectra into individual doublets allowed us to find the $E_b(Pt4f_{7/2})$ value and the contribution of different components to the experimental spectrum (Figure 5b).

Three states with $E_b(Pt4f_{7/2}) = 71.7$, 72.9, and 74.2 eV can be distinguished in the Pt4f spectra (Figure 5b). The peak with $E_b(Pt4f_{7/2}) = 71.7$ eV corresponds to metallic platinum particles Pt^0. A small increase in $E_b(Pt4f_{7/2})$ with respect to the value typical of bulk metallic platinum (71.1 eV) [18] indicates the formation of small nanoparticles [29,30] and is related to the photoemission relaxation effect [31,32]. The peak with the binding energy of 74.2 eV can be assigned to the oxidized platinum species Pt^{4+} [33,34]. The oxidized platinum species may emerge due to the oxidation of nanoparticles when samples make contact with the atmosphere during the transfer to the spectrometer. The peak with the binding energy of 72.9 eV can be attributed to the oxidized platinum species Pt^{2+} [29] as well as to the atomically dispersed platinum species [35–37]. As seen in Figure 5b, when going from CNTs to N–CNTs, a relative contribution of the platinum species with $E_b(Pt4f_{7/2}) = 72.9$ eV increases from 30 to 42%, which testifies to the formation of more dispersed or oxidized platinum species in this sample. Taking into account that this catalyst virtually does not chemisorb CO (CO/Pt = 15), it can be supposed that in N-doped catalysts, along with metallic platinum, the electron-deficient platinum is present. The electron-deficient platinum can be formed as Pt^{2+} species at the interfaces of Pt nanoparticles and N–CNTs [38] and isolated Pt ions.

According to the DFT calculations, the formation of such electron-deficient metal species may occur due to the interaction of metals with pyridinic sites of various N–CNMs [5,7–11].

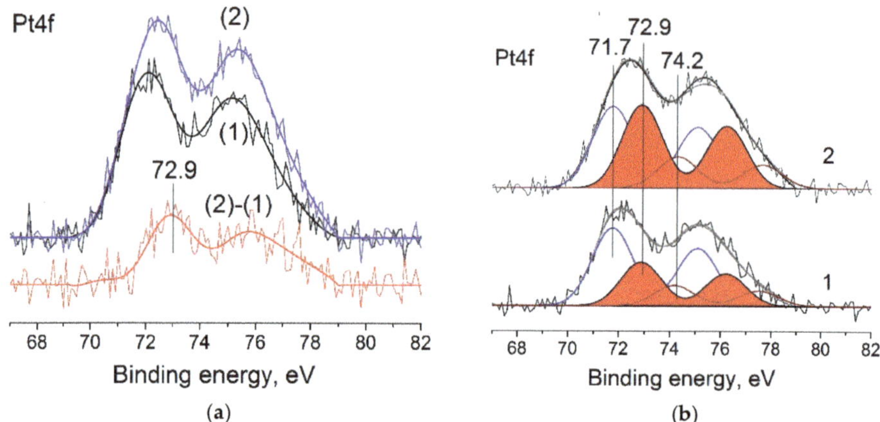

Figure 5. (**a**) Pt4f spectra and (**b**) curve-fitted Pt4f spectra of 0.2%Pt/CNTs-pr (1) and 0.2%Pt/N-CNTs-pr (2) catalysts.

To confirm the interaction of metals with certain nitrogen sites of N–CNMs, a detailed analysis of the XPS spectra of deposited metal and nitrogen is commonly used. In [39], the $E_b(Pt4f_{7/2})$ shift to a lower level was assigned to the interaction of Pt nanoparticles with nitrogen centers of the CN@Pt/CNTs catalyst. In [40], the depth of XPS analysis decreased when using the energy of monochromatized synchrotron radiation at 800 eV; this allowed the authors to observe a decrease in the intensity of the pyridinic nitrogen line and the appearance of a new line at 398.7 eV, which was assigned to the interaction of Pd with pyridinic sites of the N–C_M material. Although in [41] the XPS experiments were performed at a high energy of x-ray radiation, ca. 1500 eV, the authors observed a shift in the graphitic nitrogen peak to the high energy side by 0.2 eV upon Pt deposition on N–CNTs. It should be noted that the N–CNTs were synthesized by post-treatment of the oxidized CNTs with ammonia, and nitrogen seems to be located most likely only on the tube surface. In [8], simultaneous changes were observed in positions of the lines corresponding to both the graphitic and pyridinic nitrogen upon platinum deposition on N–CNTs, depending on the synthesis method and platinum amount; nevertheless, it was concluded that only one of the nitrogen species was responsible for stabilization.

In our study, we primarily verified that nitrogen was incorporated into the carbon's structure: this was indicated by the 0.4 eV shift of the C1s spectrum relative to the position typical of sp^2-hybride carbon structures (284.4 eV) and by broadening of the line toward high energies (Figure 6a) [42,43]. The analysis of the N1s spectra revealed that the main structural forms of nitrogen observed in N–CNMs are present in N–CNTs: pyridinic (N_{Py} 398.5 eV), pyrrolic (N_{Pyr} 399.6 eV), and graphitic (N_Q 401.0 eV) [9,43,44]. Peaks in the region of 402.5 and 404.7 eV correspond to the oxidized state and molecular nitrogen encapsulated inside the N–CNTs (Figure 6b).

A comparison of the N1s spectra for N–CNTs and 0.2%Pt/N–CNTs-pr, which is illustrated in Table 2, did not allow a reliable conclusion on the type of nitrogen involved in platinum anchoring. Investigation of samples with the platinum content of 1 wt % also did not clear up this question. All of the recorded differences were within the measurement error, although a slight decrease in the N_{Py} content upon increase in the Pt concentration in the catalysts was observed. Thus, it can be stated that the possibilities of XP spectroscopy used in this study make it possible to reliably conclude only on an increase in the relative amount of platinum with $E_b(Pt4f_{7/2})$ = 72.9 eV in N-doped catalysts. Sensitivity of the methods should be increased considerably to make a reliable and detailed interpretation of the changes that occur in the N1s spectra upon platinum deposition on N–CNTs, in which, as we have shown, nitrogen is distributed uniformly in the bulk of the material [15].

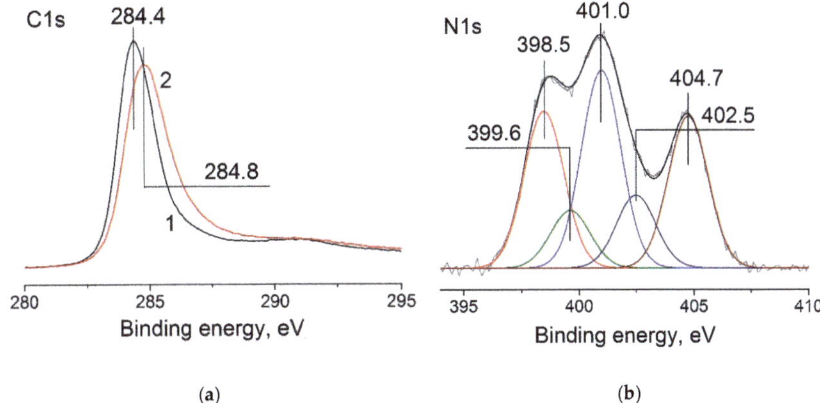

Figure 6. C1s (**a**) of 0.2%Pt/CNTs-pr (1) and 0.2%Pt/N–CNTs-pr (2) catalysts and N1s (**b**) XPS spectra of 0.2%Pt/N–CNTs-pr catalyst.

Table 2. Content of different nitrogen species according to XPS data.

Sample	N_{Py}		N_{Pyr}		N_Q		N–Ox		N_2		N
	Eb	wt %	Eb	wt %	Eb	wt %	Eb	wt %	Eb	wt %	wt %
N–CNTs	398.4	1.4	399.6	0.50	401.0	1.75	402.5	0.60	404.7	1.35	5.6
0.2%Pt/N–CNTs-pr	398.4	1.35	399.6	0.50	401.0	1.70	402.5	0.70	404.7	1.35	5.6
1%Pt/N–CNTs-pr	398.4	1.30	399.5	0.50	400.9	1.70	402.3	0.65	404.7	1.35	5.5

Finally, we compared the Pt catalysts deposited on the CNTs and bamboo-like N–CNTs with corresponding Pt catalysts deposited on herring-bone CNFs and N–CNFs. It should be noted that nitrogen-free and nitrogen-doped carbon nanotubes and nanofibers were synthesized by the same CVD method using ethylene or the standard ethylene–ammonia reaction mixture (see [19] and this work). Figure 7 shows the activity of these catalysts in the formic acid decomposition reaction. As seen in Figure 7, there was a rather good correlation between the TOF value and nitrogen content in the materials, irrespective of the essential differences in their structure related to the packing of graphene layers.

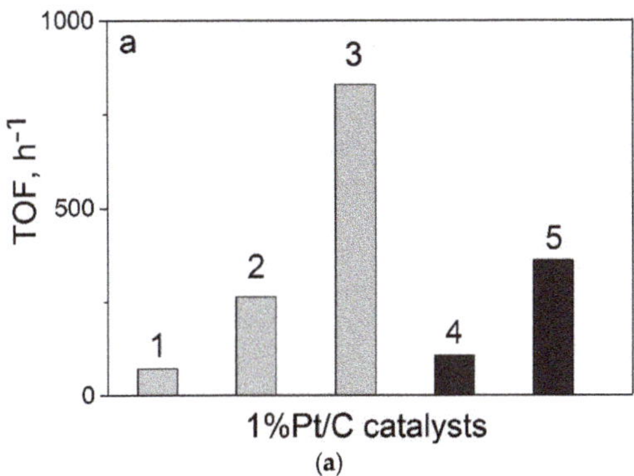

Figure 7. *Cont.*

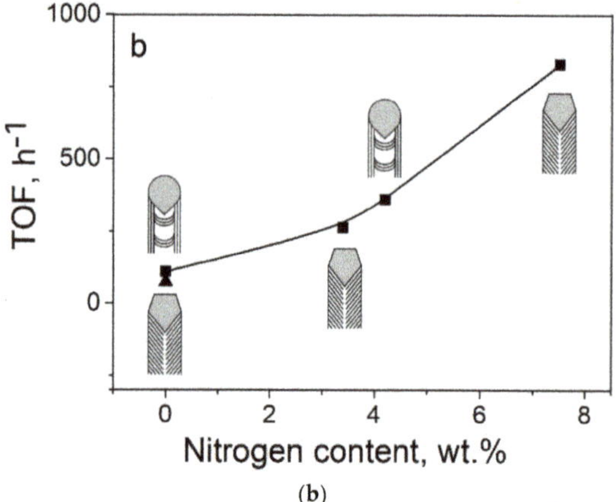

(b)

Figure 7. The TOF values at 125 °C for the 1%Pt/C catalysts synthesized by the homogeneous precipitation route depending on (**a**) carbon supports: CNFs (1), 3.4%N–CNFs (2), 7.5%N–CNFs (3), CNTs (4), 5.5%N–CNTs (5); and (**b**) nitrogen content in the catalysts (for 5.5%N–CNTs the capsulated N_2 (1.35 wt %) was excluded). TOFs for CNFs/N–CNFs-based catalysts were calculated using the data presented in [19].

Moreover, for the carbon tube- and carbon fiber-based catalysts, the decrease in the platinum content to 0.2–0.3 wt % led to the same dramatic decrease in the CO/Pt values to an extremely low level (<20%) and, on the contrary, to a substantial increase in TOF (see [5] and this work). The latter allowed us to conclude that pyridinic nitrogen also plays a key role in the formation of electron-deficient platinum in the case of N–CNTs. The amount of electron-deficient platinum depends on the content of pyridinic nitrogen, which in turn is known to be dependent on the total amount of nitrogen in N–CNTs and N–CNFs [13]. The additional arguments in favor of this conclusion are the similar values of $E_b(Pt4f_{7/2})$ for the catalysts with low platinum content, 0.2%Pt/N–CNTs-pr, and 0.3%Pt/N–CNFs, showing the high activity in the FA decomposition reaction (Figure 8).

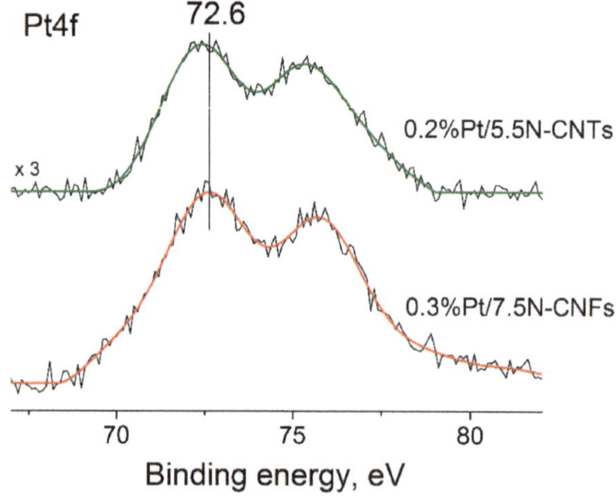

Figure 8. Pt4f spectra of the 0.3%Pt/N–CNFs-pr and 0.2%Pt/N–CNTs-pr catalysts. For 0.3%Pt/N–CNFs-pr, the data presented in [19] were used.

Indeed, metal species with high binding energies, which are typical of electron-deficient metals or single atoms stabilized by nitrogen sites, were observed in the case of N–CNMs with different structures such as bamboo-like tubes [8,9,12], carbon nanofibers with a herring-bone structure [5], porous carbon materials [6,7,45], or carbon nitride [10,46]. In the listed materials, nitrogen is present in all standard states (graphitic, pyrrolic, and pyridinic), which makes possible the locally one-type interaction of metals with nitrogen sites irrespective of their considerable structural differences. The authors of [8] explained the formation of electron-deficient platinum by the interaction of platinum with the acceptor defects containing pyridinic nitrogen and carbon vacancy. Such structural fragments of the graphene layer of C_3N_4, as shown in [10,46], also efficiently stabilizes the atomic palladium. This allows for a conclusion that, irrespective of the structural features of various N–CNMs, these materials have common properties concerning the interaction with metals; among them the formation of highly stable (Figure 1c,d) electron-deficient metal nanoparticles, isolated ions, or single atoms via the interaction with pyridinic nitrogen sites. The amount of these species depends on the content of pyridinic nitrogen, which in turn is usually determined by the total amount of nitrogen in various N–CNMs [13]. These species poorly chemisorb CO, but are highly active in the selective hydrogenation of acetylene, ammonia–borane hydrolysis, one-pot conversion of cellulose, and formic acid decomposition for the production of pure hydrogen. Thus, the possibility of controlling the size and electronic state of deposited metals makes a broad class of N–CNMs unique materials for the synthesis of efficient supported catalysts for different purposes.

4. Conclusions

The nitrogen species in N–CNTs manifest themselves as efficient anchoring sites for both Pt nanoparticles and single atoms, irrespective of the method used for Pt deposition (impregnation or homogeneous precipitation). Nitrogen in N–CNTs favors the formation of electron-deficient platinum in comparison with N-free carbon nanotubes. Moreover, it was shown that both the activity and selectivity of ionic platinum in the decomposition of formic acid were considerably higher when compared to Pt in the fully metallic state. The comparison of the results with those for Pt/N–CNFs and with the literature data for other metals and nitrogen-doped carbon supports of different structures allows us to conclude that the interaction between metal and nitrogen species has a general nature and the electron deficiency of the metal as well as the catalytic properties are mainly dependent on the concentration of pyridinic nitrogen.

Author Contributions: Conceptualization, O.P. and V.P.; Data curation, A.L. and A.B.; Investigation, O.P., A.L., L.K., A.B., O.S., V.Z., A.S., and V.S.; Methodology, O.P.; Supervision, V.P.; Writing—original draft, O.P.; Writing—review & editing, A.L., L.K., A.B., and V.P.

Funding: This work was supported by the Russian Science Foundation (Grant No. 17-73-30032).

Conflicts of Interest: The authors declare no conflicts of interest.

References

1. Grasemann, M.; Laurenczy, G. Formic acid as a hydrogen source—Recent developments and future trends. *Energy Environ. Sci.* **2012**, *5*, 8171–8181. [CrossRef]

2. Singh, A.K.; Singh, S.; Kumar, A. Hydrogen energy future with formic acid: A renewable chemical hydrogen storage system. *Catal. Sci. Tevhnol.* **2016**, *6*, 12–40. [CrossRef]

3. Zhu, J.; Holmen, A.; Chen, D. Carbon nanomaterials in catalysis: Proton affinity, chemical and electronic properties, and their catalytic consequences. *ChemCatChem* **2013**, *5*, 378–401. [CrossRef]

4. Su, D.S.; Perathoner, S.; Centi, G. Nanocarbons for the development of advanced catalysts. *Chem. Rev.* **2013**, *113*, 5782–5816. [CrossRef]

5. Bulushev, D.A.; Zacharska, M.; Lisitsyn, A.S.; Podyacheva, O.Y.; Hage, F.S.; Ramasse, Q.M.; Bangert, U.; Bulusheva, L.G. Single atoms of Pt-group metals stabilized by N-doped carbon nanofibers for efficient hydrogen production from formic acid. *ACS Catal.* **2016**, *6*, 3442–3451. [CrossRef]

6. Tang, C.; Surkus, A.; Chen, F.; Pohl, M.; Agostini, G.; Schneider, M.; Junge, H.; Beller, M. A stable nanocobalt catalyst with highly dispersed CoNx active sites for the selective dehydrogenation of formic acid. *Angew. Chem. Int. Ed.* **2017**, *56*, 16616–16620. [CrossRef] [PubMed]

7. Zacharska, M.; Bulusheva, L.G.; Lisitsyn, A.S.; Beloshapkin, S.; Guo, Y.; Chuvilin, A.L.; Shlyakhova, E.V.; Podyacheva, O.Y.; Leahy, J.L.; Okotrub, A.V.; et al. Factors influencing the performance of Pd/C catalysts in the green production of hydrogen from formic acid. *ChemSusChem* **2017**, *10*, 720–730. [CrossRef]

8. Ning, X.; Li, Y.; Dong, B.; Wang, H.; Yu, H.; Peng, F.; Yang, Y. Electron transfer dependent catalysis of Pt on N-doped carbon nanotubes: Effects of synthesis method on metal-support interaction. *J. Catal.* **2017**, *348*, 100–109. [CrossRef]

9. Arrigo, R.; Schuster, M.E.; Xie, Z.; Yi, Y.; Wowsnick, G.; Sun, L.L.; Hermann, K.E.; Friedrich, M.; Kast, P.; Hävecker, M.; et al. Nature of the N-Pd interaction in nitrogen-doped carbon nanotube catalysts. *ACS Catal.* **2015**, *5*, 2740–2753. [CrossRef]

10. Huang, X.; Xia, Y.; Cao, Y.; Zheng, X.; Pan, H.; Zhu, J.; Ma, C.; Wang, H.; Li, J.; You, R.; et al. Enhancing both selectivity and coking-resistance of a single-atom Pd1/C3N4 catalyst for acetylene hydrogenation. *Nano Res.* **2017**, *10*, 1302–1312. [CrossRef]

11. Vorobyeva, E.; Chen, Z.; Mitchell, S.; Leary, R.K.; Midgley, P.; Thomas, J.M.; Hauert, R.; Fako, E.; López, N.; Pérez-Ramírez, J. Tailoring the framework composition of carbon nitride to improve the catalytic efficiency of the stabilised palladium atoms. *J. Mater. Chem. A* **2017**, *5*, 16393–16403. [CrossRef]

12. Podyacheva, O.Y.; Bulushev, D.A.; Suboch, A.N.; Svintsitskiy, D.A.; Lisitsyn, A.S.; Modin, E.; Chuvilin, A.; Gerasimov, E.Y.; Sobolev, V.I.; Parmon, V.N. Highly stable single-atom catalyst with ionic Pd active sites supported on N-doped carbon nanotubes for formic acid decomposition. *ChemSusChem* **2018**, *11*, 3724–3727. [CrossRef]

13. Podyacheva, O.Y.; Cherepanova, S.V.; Romanenko, A.I.; Kibis, L.S.; Svintsitskiy, D.A.; Boronin, A.I.; Stonkus, O.A.; Suboch, A.N.; Puzynin, A.V.; Ismagilov, Z.R. Nitrogen doped carbon nanotubes and nanofibers: Composition, structure, electrical conductivity and capacity properties. *Carbon* **2017**, *122*, 475–483. [CrossRef]

14. Shalagina, A.E.; Ismagilov, Z.R.; Podyacheva, O.Y.; Kvon, R.I.; Ushakov, V.A. Synthesis of nitrogen-containing carbon nanofibers by catalytic decomposition of ethylene/ammonia mixture. *Carbon* **2007**, *45*, 1808–1820. [CrossRef]

15. Svintsitskiy, D.A.; Kibis, L.S.; Smirnov, D.A.; Suboch, A.N.; Stonkus, O.A.; Podyacheva, O.Y.; Boronin, A.I.; Ismagilov, Z.R. Spectroscopic study of nitrogen distribution in N-doped carbon nanotubes and nanofibers synthesized by catalytic ethylene-ammonia decomposition. *Appl. Surf. Sci.* **2018**, *435*, 1273–1284. [CrossRef]

16. Suboch, A.N.; Cherepanova, S.V.; Kibis, L.S.; Svintsitskiy, D.A.; Stonkus, O.A.; Boronin, A.I.; Chesnokov, V.V.; Romanenko, A.I.; Ismagilov, Z.R.; Podyacheva, O.Y. Observation of the superstructural diffraction peak in the nitrogen doped carbon nanotubes: Simulation of the structure. *Fullerenes Nanotub. Carbon Nanostruct.* **2016**, *24*, 520–530. [CrossRef]

17. Evtushok, V.Y.; Suboch, A.N.; Podyacheva, O.Y.; Stonkus, O.A.; Zaikovskii, V.I.; Chesalov, Y.A.; Kibis, L.S.; Kholdeeva, O.A. Highly efficient catalysts based on divanadium-substituted polyoxometalate and N-doped carbon nanotubes for selective oxidation of alkylphenols. *ACS Catal.* **2018**, *8*, 1297–1307. [CrossRef]

18. Briggs, D.; Seah, M.P. (Eds.) *Practical Surface Analysis by Auger and X-ray Photoelectron Spectroscopy*; Wiley: New York, NY, USA, 1983.

19. Jia, L.; Bulushev, D.A.; Podyacheva, O.Y.; Boronin, A.I.; Kibis, L.S.; Gerasimov, E.Y.; Beloshapkin, S.; Seryak, I.A.; Ismagilov, Z.R.; Ross, J.R.H. Pt nanoclusters stabilized by N-doped carbon nanofibers for hydrogen production from formic acid. *J. Catal.* **2013**, *307*, 94–102. [CrossRef]

20. Rocha, R.P.; Restivo, J.; Sousa, J.P.S.; Órfão, J.J.M.; Pereira, M.F.R.; Figueiredo, J.L. Nitrogen-doped carbon xerogels as catalysts for advanced oxidation processes. *Catal. Today* **2015**, *241*, 73–79. [CrossRef]

21. Luo, J.; Peng, F.; Wang, H.; Yu, H. Enhancing the catalytic activity of carbon nanotubes by nitrogen doping in the selective liquid phase oxidation of benzyl alcohol. *Catal. Commun.* **2013**, *39*, 44–49. [CrossRef]

22. Chizari, K.; Deneuve, A.; Ersen, O.; Florea, I.; Liu, Y.; Edouard, D.; Janowska, I.; Begin, D.; Pham-Huu, C. Nitrogen-doped carbon nanotubes as a highly active metal-free catalyst for selective oxidation. *ChemSusChem* **2012**, *5*, 102–108. [CrossRef] [PubMed]

23. Zhou, K.; Li, B.; Zhang, Q.; Huang, J.Q.; Tian, G.L.; Jia, J.C.; Zhao, M.Q.; Luo, G.H.; Su, D.S.; Wei, F. The catalytic pathways of hydrohalogenation over metal-free nitrogen-doped carbon nanotubes. *ChemSusChem* **2014**, *7*, 723–728. [CrossRef] [PubMed]

24. Gao, Y.; Hu, G.; Zhong, J.; Shi, Z.; Zhu, Y.; Su, D.S.; Wang, J.; Bao, X.; Ma, D. Nitrogen-doped sp2-hybridized carbon as a superior catalyst for selective oxidation. *Angew. Chem. Int. Ed.* **2013**, *52*, 2109–2113. [CrossRef] [PubMed]

25. Zhang, Y.; Zhang, J.; Su, D.S. Substitutional doping of carbon nanotubes with heteroatoms and their chemical applications. *ChemSusChem* **2014**, *7*, 1240–1250. [CrossRef] [PubMed]

26. Podyacheva, O.Y.; Ismagilov, Z.R. Nitrogen-doped carbon nanomaterials: To the mechanism of growth, electrical conductivity and application in catalysis. *Catal. Today* **2015**, *249*, 12–22. [CrossRef]

27. Kaprielova, K.M.; Yakovina, O.A.; Ovchinnikov, I.I.; Koscheev, S.V.; Lisitsyn, A.S. Preparation of platinum-on-carbon catalysts via hydrolytic deposition: Factors influencing the deposition and catalytic properties. *Appl. Catal. A* **2012**, *449*, 203–214. [CrossRef]

28. Kaprielova, K.M.; Ovchinnikov, I.I.; Yakovina, O.A.; Lisitsyn, A.S. Synthesis of Pt/C catalysts through reductive deposition: Ways of tuning catalytic properties. *ChemCatChem* **2013**, *5*, 2015–2024. [CrossRef]

29. Vovk, E.I.; Kalinkin, A.V.; Smirnov, M.Y.; Klembovskii, I.O.; Bukhtiyarov, V.I. XPS study of stability and reactivity of oxidized Pt nanoparticles supported on TiO$_2$. *J. Phys. Chem. C* **2017**, *121*, 17297–17304. [CrossRef]

30. Podyacheva, O.Y.; Ismagilov, Z.R.; Boronin, A.I.; Kibis, L.S.; Slavinskaya, E.M.; Noskov, A.S.; Shikina, N.V.; Ushakov, V.A.; Ischenko, A.V. Platinum nanoparticles supported on nitrogen-containing carbon nanofibers. *Catal. Today* **2012**, *186*, 42–47. [CrossRef]

31. Mason, M.G. Electronic structure of supported small metal clusters. *Phys. Rev. B* **1983**, *27*, 748–762. [CrossRef]

32. Wertheim, G.K.; DiCenzo, S.B.; Buchanan, D.N.E.; Bennett, P.A. Core electron binding energy shifts in metal clusters: Tin on amorphous carbon. *Solid State Commun.* **1985**, *53*, 377–381. [CrossRef]

33. Ono, L.K.; Yuan, B.; Heinrich, H.; Cuenya, B.R. Formation and thermal stability of platinum oxides on size-selected platinum nanoparticles: Support effects. *J. Phys. Chem. C* **2010**, *114*, 22119–22133. [CrossRef]

34. Svintsitskiy, D.A.; Kibis, L.S.; Stadnichenko, A.I.; Koscheev, S.V.; Zaikovskii, V.I.; Boronin, A.I. Highly oxidized platinum nanoparticles prepared through radio-frequency sputtering: Thermal stability and reaction probability towards CO. *ChemPhysChem* **2015**, *16*, 3318–3324. [CrossRef] [PubMed]

35. Kim, Y.T.; Ohshima, K.; Higashimine, K.; Uruga, T.; Takata, M.; Suematsu, H.; Mitani, T. Fine size control of platinum on carbon nanotubes: From single atoms to clusters. *Angew. Chem. Int. Ed.* **2006**, *45*, 407–411. [CrossRef]

36. Eberhardt, W.; Fayet, P.; Cox, D.; Fu, Z.; Kaldor, A.; Sherwood, R.; Sondericker, D. Core level photoemission from monosize mass selected pt clusters deposited on SiO$_2$ and amorphous carbon. *Phys. Scr.* **1990**, *41*, 892–895. [CrossRef]

37. Kawasaki, H.; Yamamoto, H.; Fujimori, H.; Arakawa, R.; Inada, M.; Iwasaki, Y. Surfactant-free solution synthesis of fluorescent platinum subnanoclusters. *Chem. Commun.* **2010**, *46*, 3759–3761. [CrossRef]

38. Arrigo, R.; Schuster, M.E.; Abate, S.; Giorgianni, G.; Centi, G.; Perathoner, S.; Wrabetz, S.; Pfeifer, V.; Antonietti, M.; Schlogl, R. Pd Supported on carbon nitride boosts the direct hydrogen peroxide synthesis. *ACS Catal.* **2016**, *6*, 6959–6966. [CrossRef]

39. Xia, L.; Li, D.; Long, J.; Huang, F.; Yang, L.; Guo, Y.; Jia, Z.; Xiao, J.; Liu, H. N-Doped graphene confined Pt nanoparticles for efficient semi-hydrogenation of phenylacetylene. *Carbon* **2019**, *145*, 47–52. [CrossRef]

40. Bulushev, D.A.; Zacharska, M.; Shlyakhova, E.V.; Chuvilin, A.L.; Guo, Y.; Beloshapkin, S.; Okotrub, A.V.; Bulusheva, L.G. Single isolated Pd2+ cations supported on N-doped carbon as active sites for hydrogen production from formic acid decomposition. *ACS Catal.* **2016**, *6*, 681–691. [CrossRef]

41. Shi, W.; Zhang, B.; Lin, Y.; Wang, Q.; Zhang, Q.; Su, D.S. Enhanced chemoselective hydrogenation through tuning the interaction between Pt nanoparticles and carbon supports: Insights from identical location transmission electron microscopy and X-ray photoelectron spectroscopy. *ACS Catal.* **2016**, *6*, 7844–7854. [CrossRef]

42. Jang, J.W.; Lee, C.E.; Lyu, S.C.; Lee, T.J.; Lee, C.J. Structural study of nitrogen-doping effects in bamboo-shaped multiwalled carbon nanotubes. *Appl. Phys. Lett.* **2004**, *84*, 2877–2879. [CrossRef]

43. Susi, T.; Pichler, T.; Ayala, P. X-Ray photoelectron spectroscopy of graphitic carbon nanomaterials doped with heteroatoms. *Beilstein J. Nanotechnol.* **2015**, *6*, 177–192. [CrossRef] [PubMed]

Energies **2019**, *12*, 3976

44. Roldán, L.; Armenise, S.; Marco, Y.; García-Bordejé, E. Control of nitrogen insertion during the growth of nitrogen-containing carbon nanofibers on cordierite monolith walls. *Phys. Chem. Chem. Phys.* **2012**, *14*, 3568–3575. [CrossRef] [PubMed]

45. Liu, W.; Chen, Y.; Qi, H.; Zhang, L.; Yan, W.; Liu, X.; Yang, X.; Miao, S.; Wang, W.; Liu, C.; et al. A durable nickel single-atom catalyst for hydrogenation reactions and cellulose valorization under harsh conditions. *Angew. Chem. Int. Ed.* **2018**, *57*, 7071–7075. [CrossRef]

46. Vile, G.; Albani, D.; Nachtegaal, M.; Chen, Z.; Dontsova, D.; Antonietti, M.; López, N.; Perez-Ramírez, J. A stable single-site palladium catalyst for hydrogenations. *Angew. Chem. Int. Ed.* **2015**, *54*, 11265–11269. [CrossRef]

Article

Production of Hydrogen-Rich Gas by Formic Acid Decomposition over CuO-CeO$_2$/γ-Al$_2$O$_3$ Catalyst

Alexey Pechenkin [1,2], Sukhe Badmaev [1,2,*], Vladimir Belyaev [1] and Vladimir Sobyanin [1]

[1] Boreskov Institute of Catalysis, Novosibirsk 630090, Russia; pechenkin@catalysis.ru (A.P.); belyaev@catalysis.ru (V.B.); sobyanin@catalysis.ru (V.S.)

[2] Department of Natural Sciences, Novosibirsk State University, Novosibirsk 630090, Russia

* Correspondence: sukhe@catalysis.ru

Received: 29 August 2019; Accepted: 12 September 2019; Published: 19 September 2019

Abstract: Formic acid decomposition to H$_2$-rich gas was investigated over a CuO-CeO$_2$/γ-Al$_2$O$_3$ catalyst. The catalyst was characterized by XRD, HR TEM and EDX methods. A 100% conversion of formic acid was observed over the copper-ceria catalyst under ambient pressure, at 200–300 °C, N$_2$:HCOOH = 75:25 vol.% and flow rate 3500–35,000 h^{-1} with H$_2$ yield of 98%, wherein outlet CO concentration is below the equilibrium data (<0.5 vol.%). The copper-ceria catalyst proved to be promising for multifuel processor application, and the H$_2$ generation from dimethoxymethane, methanol, dimethyl ether and formic acid on the same catalyst for fuel cell supply.

Keywords: formic acid decomposition; hydrogen production; CuO-CeO$_2$/γ-Al$_2$O$_3$; multifuel processor; copper catalyst; oxygenates; fuel cell

1. Introduction

Growing interest in, and demand for, high temperature proton exchange membrane fuel cells (HT PEM FC) has evidently manifested itself during the past decade due to their high tolerance to fuel impurities compared to the low temperature (LT) PEM FC [1–3]. For instance, the HT PEM FC can be fueled by hydrogen-rich gas containing up to 3 vol.% of CO [1]. Analysis of current literature shows that gas mixture with this composition can be produced by catalytic conversion of oxygenates such as methanol [4–6], dimethyl ether (DME) [7–9], dimethoxymethane (DMM) [10–14] and formic acid (FA) [15–23]. The obtained H$_2$-rich gas can be directly used as fuel without complex CO removal processes. This makes the HT PEM FC-based power units more compact and simple. The typical HT PEM FC-based power unit scheme is shown in Figure 1.

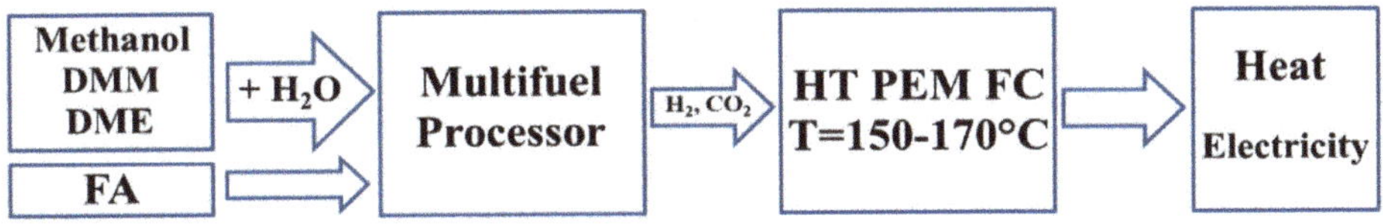

Figure 1. Scheme of power unit based on high temperature proton exchange membrane fuel cells fueled by oxygenates.

In our previous work [12,13], effective bifunctional CuO-CeO$_2$/γ-Al$_2$O$_3$ catalysts have been proposed for methanol, dimethyl ether and dimethoxymethane steam reforming (SR) to H$_2$-rich gas showing great promise for a multifuel processor approach. The catalysts comprised of acidic and Cu-based sites and provided complete methanol/DME/DMM conversion and a H$_2$ production rate of 15 L/(g$_{cat}$·h) at 300–370 °C.

Supported Cu-based catalysts proved highly active and selective for formic acid (FA) decomposition (Reaction (1)) [20]. In particular, the reaction in this work was studied over carbon-supported copper catalysts doped with nitrogen atoms (Cu/N-PCN). The catalysts provided complete conversion of FA at temperatures >270 °C and high hydrogen productivity (up to 97.4%).

$$HCOOH = CO_2 + H_2 \tag{1}$$

Reaction (1) can be accompanied by side Reactions (2) and (3) yielding CO and water:

$$HCOOH = CO + H_2O \tag{2}$$

$$CO_2 + H_2 = CO + H_2O \tag{3}$$

The present work reports the performance of the CuO-CeO_2/γ-Al_2O_3 catalysts in FA decomposition to H_2-rich gas to be used for HT PEM FC. The effect of some reaction conditions on the catalyst's performance was studied. The catalyst was characterized by XRD, HRTEM and EDX techniques. In order to evaluate the catalyst's possibility for multifuel processor applications, we compared CuO-CeO_2/γ-Al_2O_3 catalytic properties in methanol, dimethyl ether and dimethoxymethane steam reforming and FA decomposition reactions.

2. Materials and Methods

The catalyst of composition 10 wt.% CuO-5 wt.% CeO_2/γ-Al_2O_3, which showed the best performance in methanol, DME and DMM SR reactions in our earlier study [13], was used in the FA decomposition experiments. It was prepared by the impregnation method reported in details elsewhere [13]. Samples for transmission electron microscopy (TEM) were prepared by carefully applying the catalyst powder on a carbon film on a nickel grid. These samples were investigated by conventional and high-resolution (HR) TEM and energy-dispersive X-ray (EDX) chemical microanalysis, using an electron microscope JEM 2010 (JEOL Ltd., Tokyo, Japan) at operation voltage 200 kV and lattice-fringe resolution 0.14 nm. The microscope was equipped with an EDX unit (EDAX Co, Mahwah, NJ, USA) for the local elemental analysis (the energy resolution was 130 eV).

X-ray diffraction (XRD) data were collected using a URD-63 diffractometer with a CuK_α source (graphite monochromator). The measurements were taken within 2θ scanning area = $20°$–$80°$ with a step of $0.02°$ (2θ) and an accumulation time of 1.0 s. The diffraction patterns were processed using the PowderCell 2.4 software package. The identification of the crystalline phases was performed using the JCPDS international diffraction database.

Experiments on FA decomposition were carried out in a flow U-shaped reactor (inner diameter 4 mm) at 120–320 °C under ambient pressure. The temperature of the catalytic bed was measured using a chromel/alumel thermocouple placed in the middle of this bed. Before reaction, the copper-ceria catalyst was pre-reduced in situ at 300 °C in a mixture of hydrogen (5 vol.%) diluted with N_2 at a total flow rate of 3000 mL/h during 1 h. After that, the reactor was cooled to 120–200 °C in a flowing mixture of hydrogen and nitrogen and then the gas mixture was switched to (vol.%): 25 HCOOH and 75 N_2. The total gas hourly space velocity (GHSV) was 3500–35,000 h^{-1}.

FA was fed into the reactor by passing N_2 (99.999%) through a glass saturator filled with formic acid (JSC Reahim, 99%) maintained at 60 °C. N_2 and H_2 were introduced to the reactor by gas flow-mass meters (Bronkhorst). The inlet and outlet concentrations of H_2, CO, CO_2 and H_2O were recorded using gas chromatography (Chromos-1000, Chromos Engineering, Dzerzhinsk, Russia) equipped with thermal conductivity and flame ionization detectors (Porapak T and CaA molecular sieves columns). The minimum concentration of CO, CO_2, H_2 and H_2O that could be determined using this method was $5 \cdot 10^{-3}$ vol.%. During the experiments, the carbon balance was controlled with an accuracy of ±3%.

Conversion of HCOOH, H_2 selectivity, hydrogen yield and productivity were calculated by the following formulas:

$$X_{HCOOH}(\%) = \frac{C_{HCOOH}^{in} - C_{HCOOH}^{out} \times \frac{C_{N_2}^{in}}{C_{N_2}^{out}}}{C_{HCOOH}^{in}} \times 100 \tag{4}$$

$$S_{H_2}(\%) = \frac{C_{CO_2}^{out}}{C_{CO_2}^{out} + C_{CO}^{out}} \times 100 \tag{5}$$

$$Y_{H_2}(\%) = \frac{S_{H_2} \times X_{HCOOH}}{100} \tag{6}$$

$$W_{H_2}\left(\frac{L}{h \cdot g_{cat}}\right) = \frac{F \times C_{H_2}^{out} \times \frac{C_{N_2}^{in}}{C_{N_2}^{out}}}{100 \times m_{cat}} \tag{7}$$

where C^{in}, C^{out} are the concentrations before and after reactor, respectively; F is the flow rate of the reaction mixture (L/h); m_{cat} is the catalyst weight (g).

3. Results

3.1. Catalyst Characterization

Table 1 presents the data obtained from XRD patterns of fresh and used CuO-CeO_2/γ-Al_2O_3 catalysts. It shows that the fresh catalyst contains phases of cerium dioxide and γ-Al_2O_3. A strong decrease in the value of cerium dioxide unit cell parameter to 5.390 Å, as compared to the standard value of 5.411 Å, was observed. This may be due to the insertion of copper ions with a smaller ionic radius and a lesser degree of oxidation in the structure of cerium dioxide (r (Cu^{2+}) = 0.76 Å, r (Ce^{4+}) = 1.01 Å). CeO_2 is in a highly dispersed state; its CSR size is 40 Å. The phases of crystallized copper species were not detected. Apparently, this fact indicates that copper oxide is in a form of either well dispersed or amorphous species on γ-Al_2O_3 surface. For the used catalyst, we can see a slight decrease in the CeO_2 lattice parameter to a value of 5.395 Å. This fact most likely indicates the possible formation of a mixed oxide of copper and cerium [24–26]. The phase of highly dispersed metallic copper is also observed in the used catalyst. Most likely, this is due to the fact that part of the copper ions leave the structure of cerium oxide during the reaction.

Table 1. XRD data for the CuO-CeO_2/γ-Al_2O_3 catalysts.

Catalyst	Phase Composition	Unit Cell Parameters, Å	CSR (Å)
Fresh	γ-Al_2O_3 CeO_2	7.918 5.390	50 40
Used	γ-Al_2O_3 CeO_2 Cu	7.918 5.395 n.d.	50 30 highly dispersed

Figure 2 shows the TEM image of the fresh CuO-CeO_2/γ-Al_2O_3 catalyst and the HR TEM image and EDX spectrum of the marked area. In the TEM image, we observed an agglomerate of about 50 nm in size (the marked area). The HR TEM image and EDX spectrum of this area were recorded. The EDX spectrum shows that this agglomerate consists of Cu and Ce and an atomic ratio Cu:Ce = 70:30 at.%. In the high resolution TEM picture of the agglomerate, cerium dioxide particle (interplanar distances d(111) = 0.312 and d(200) = 0.271 nm, not shown in figure) was observed. We were unable to detect copper particles in the HR TEM image because of their low contrast. According to XRD data, copper oxide phases were not observed in the fresh catalyst, so it can be concluded that copper oxide particles were in a highly dispersed state.

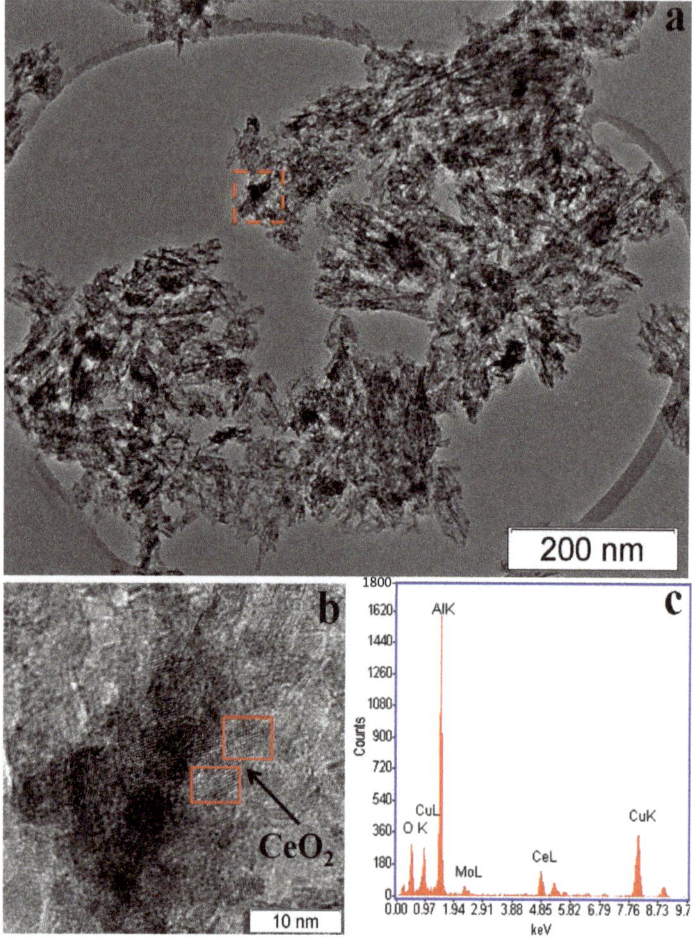

Figure 2. TEM image (**a**) of the fresh CuO-CeO$_2$/γ-Al$_2$O$_3$ catalyst and the HR TEM (**b**) and EDX spectra (**c**) of the marked area.

Figure 3 shows the TEM, HR TEM images and EDX spectrum of the CuO-CeO$_2$/γ-Al$_2$O$_3$ catalyst after reaction. It shows that, similar to the fresh catalyst, the used catalyst contains copper-ceria agglomerates of about 30–50 nm in size and an atomic ratio Cu:Ce = 60:40 at.%. In the HR TEM image, we observed CeO$_2$ particles and no copper particles, as in the case of the fresh catalyst. However, according to XRD data, the used catalyst contains the phase of highly dispersed metallic copper. We believe that this phase refers to particles outside the agglomerates. These particles obviously suffer sintering during long-term operation of the catalyst and therefore lose activity. At the same time, the active sites of the catalyst are particles of copper stabilized by cerium oxide which do not undergo changes during the formic acid decomposition reaction.

So, the CuO-CeO$_2$/γ-Al$_2$O$_3$ catalyst surface contains the agglomerates of 30–50 nm in size which consist of mixed copper-cerium oxide (solid solution) particles of 5–7 nm in size, and some fine dispersed copper. We suggest that these agglomerated mixed copper-cerium oxide particles are the active coppers sites of the catalyst, which is in agreement with our previous work [13].

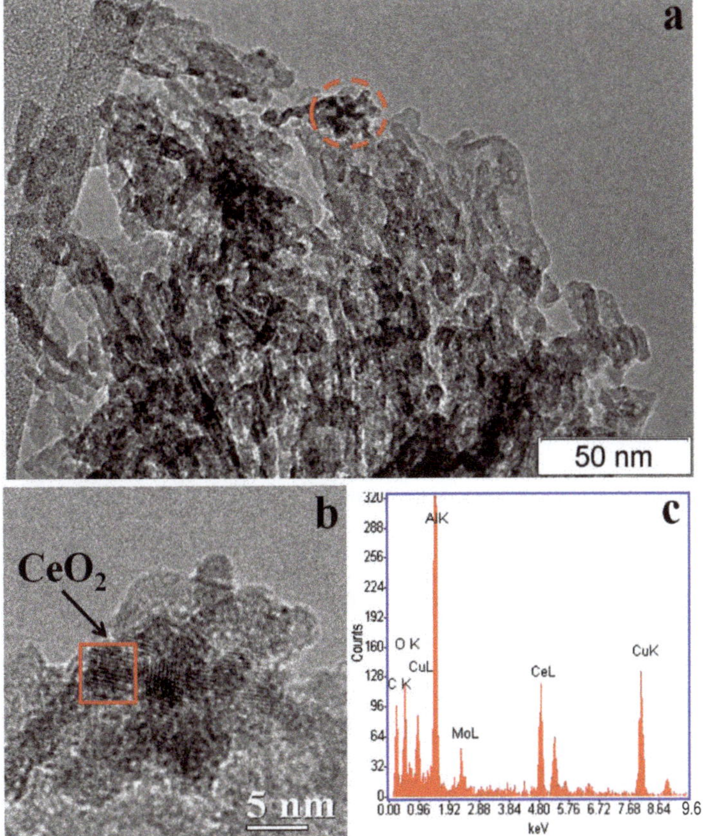

Figure 3. TEM image (**a**) of the used CuO-CeO$_2$/γ-Al$_2$O$_3$ catalyst and the HR TEM (**b**) and EDX spectra (**c**) of the marked area.

3.2. Catalytic Performance of CuO-CeO$_2$/γ-Al$_2$O$_3$ in FA Decomposition

Figure 4 demonstrates dependencies of FA conversion and outlet H$_2$, CO$_2$, CO concentrations at FA decomposition over 10%Cu-5%CeO$_2$/γ-Al$_2$O$_3$ catalyst at GHSV of 3500 h^{-1} (Figure 4a) and 35,000 h^{-1} (Figure 4b) on temperature. Equilibrium values are shown by dashed lines. The equilibrium compositions were calculated assuming equilibrium mixtures contained only H$_2$, H$_2$O, CO$_2$ and CO. The equilibrium and experimental values of H$_2$O are not shown because they coincide with respective values of CO, as well as the equilibrium values of CO$_2$ which are nearly similar to those of H$_2$.

The products of formic acid decomposition were H$_2$, H$_2$O, CO$_2$ and CO; no other products were observed. As shown in Figure 4, FA conversion rises with increasing temperature and gets 100% at 200 °C, GHSV = 3500 h^{-1}, and at 300 °C, GHSV = 35,000 h^{-1}. The product concentrations also increase with increasing temperature. In the case of GHSV = 3500 h^{-1}, at temperatures above 200 °C, the H$_2$ and CO$_2$ concentrations are above their equilibrium composition, whereas the CO concentration is below the equilibrium value. This is because the RWGS Reaction (3) does not reach equilibrium during the experiment. Such behavior is typical for this catalyst; similar results were obtained in the steam reforming of methanol, DME and DMM [13]. As GHSV increased to 35,000 h^{-1}, the temperature dependencies of FA conversion and product concentrations shifted to a higher temperature region.

Complete FA conversion was reached at 300 °C, whereas the CO concentration remained constant and stayed below 0.5 vol.%.

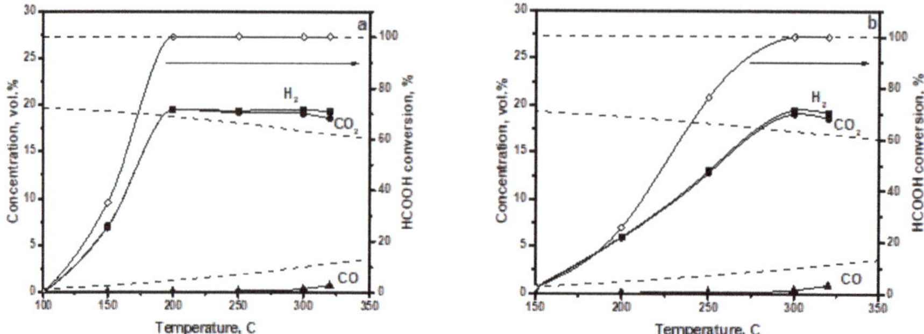

Figure 4. Effect of temperature on formic acid conversion and H_2, CO and CO_2 outlet concentrations in formic acid decomposition over CuO-CeO$_2$/γ-Al$_2$O$_3$ catalyst at gas hour space velocity = 3500 h^{-1} (**a**) and gas hour space velocity = 35,000 h^{-1} (**b**). Reaction conditions: p = 1 atm, inlet composition, vol.%: N$_2$:HCOOH = 75:25. Solid lines—experiment, dotted lines—thermodynamic equilibrium values.

Figure 5 illustrates how the hydrogen yield and CO concentration depended on the reaction mixture feed rate at FA decomposition over CuO-CeO$_2$/γ-Al$_2$O$_3$ at 300 °C. It shows that, even with ten-fold increase in GHSV, the hydrogen yield and CO concentration remained almost unchanged and amount to 98% and 0.4 vol.%, respectively.

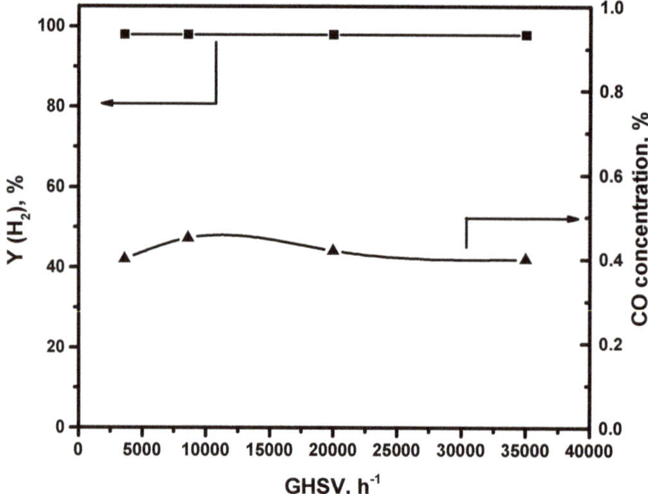

Figure 5. Effect of GHSV on H_2 yield (■) and outlet CO concentration (▲) over CuO-CeO$_2$/γ-Al$_2$O$_3$ catalyst for HCOOH decomposition. Reaction conditions: p = 1 atm, T = 300 °C, X (HCOOH) = 100%. Inlet composition, vol.%: N$_2$:HCOOH = 75:25.

Figure 6 illustrates the effect of time on stream on the FA conversion and the outlet concentrations of H_2 and CO in FA decomposition over CuO-CeO$_2$/γ-Al$_2$O$_3$. The experiment was carried out at 250 °C and GHSV = 35,000 h^{-1}. It showed that during 8 h on stream, the catalyst was stable and supported constant values of FA conversion and H_2 and CO concentrations. We tend to explain this catalyst stability by the fact that the copper particles on the CuO-CeO$_2$/γ-Al$_2$O$_3$ surface were stabilized by ceria and thus protected against sintering.

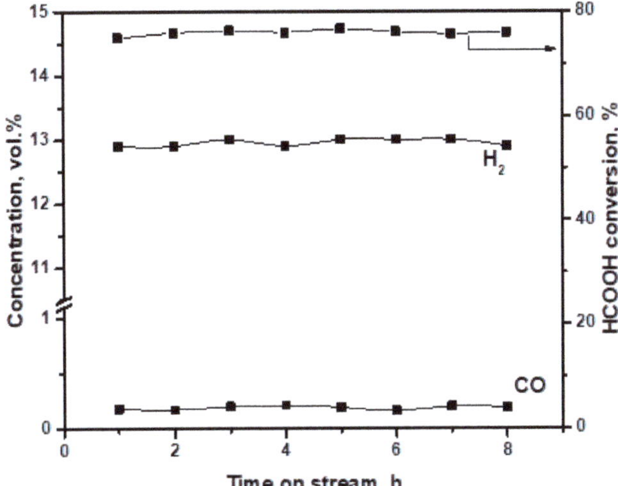

Figure 6. Effect of time on stream on CO and hydrogen outlet concentrations in FA decomposition over CuO-CeO$_2$/γ-Al$_2$O$_3$ catalyst. Reaction conditions: p = 1 atm, T = 250 °C, GHSV = 35,000 h^{-1}; Inlet composition, vol.%: N$_2$:HCOOH = 75:25.

So, the results obtained (Figures 4–6) show that complete conversion of FA over 10% CuO-5% CeO$_2$/γ-Al$_2$O$_3$ was reached at temperatures 200–300 °C, yielding H$_2$ and CO$_2$ as the main products. The CO concentration was below 0.5 vol.%. The catalyst provided high hydrogen yield (~100%) in a wide range of GHSV = 3500–35,000 h^{-1} at temperatures 200–300 °C.

In this work, the catalyst properties were studied using reaction mixture diluted with nitrogen. Note again that nitrogen was used as an internal standard to provide correct determination of the reaction parameters. Clearly, hydrogen-rich gas mixture for FC feeding should be free of nitrogen dilution. Our calculations for nitrogen-free reaction mixture showed that FA decomposition over 10% CuO-5% CeO$_2$/γ-Al$_2$O$_3$ produces hydrogen-rich gas of composition (vol.%): 48 H$_2$, 48 CO$_2$, 2 CO, 2 H$_2$O that is appropriate for direct feeding HT PEM FC.

Table 2 compares the results on vapor phase FA decomposition over CuO-CeO$_2$/γ-Al$_2$O$_3$ and other efficient catalysts: Cu/N-doped carbon [20]; 1% Au/SiO$_2$ [21]; and Ir-, Pt-, Rh-, Pd-/C [22]. All data refer to complete FA conversion. Note that H$_2$ and CO$_2$ were the main products of FA decomposition over these catalysts. According to the cited data, the product distribution insignificantly depended on the reaction conditions, and the maximum yield of H$_2$ (92–99.5 %) was observed at 200–300 °C for all catalysts.

Table 2. Comparison of catalyst activities in formic acid decomposition.

| Catalysts | T, °C | Reaction Condition | | Y (H$_2$), % | Refs |
		HCOOH: Inert vol.%:vol.%	FA Flow Rate, h^{-1}		
10% CuO-5% CeO$_2$/γ-Al$_2$O$_3$	200	25:75	875	98	This work
	300	25:75	8750	98	
Cu/C	270	5:95	3150 [1]	97.4	[20]
1% Au/SiO$_2$	250	7:93	560	99.5	[21]
2% Ir/C	200	5:95	400	99.5	[22]
2% Pt/C				98.6	
2% Rh/C				97	
2% Pd/C				91.7	

[1]—data calculated per gram catalyst 3150 mL/(h g$_{cat}$).

Thus, the literature data [20] and results of the present work prove that Cu-based catalysts are effective in the reaction of FA decomposition and keep up with catalysts containing the VIII group metals.

As mentioned above (see Introduction), the CuO-CeO$_2$/γ-Al$_2$O$_3$ catalyst was effective in dimethoxymethane, dimethyl ether, methanol SR reactions [12,13]. It seems reasonable to compare the catalytic properties of CuO-CeO$_2$/γ-Al$_2$O$_3$ catalyst in dimethoxymethane, dimethyl ether, methanol SR and FA decomposition reactions in order to extend the scope of multifuel processor applications.

Table 3 presents the following data: temperature of 100% conversion of dimethoxymethane, dimethyl ether, methanol and FA to H$_2$-rich gas; H$_2$ yield; outlet CO concentration in the produced H$_2$-rich gas; and H$_2$ productivity. Complete dimethoxymethane, dimethyl ether, methanol and FA conversion was attained at 300–370 °C. Regardless of the raw material, the catalyst yielded H$_2$-rich gas containing low outlet CO concentration (≤1 vol.%), which can be used directly for HT PEM FC feeding [1–3]. H$_2$ productivity of the CuO-CeO$_2$/γ-Al$_2$O$_3$ catalyst (Table 3) in FA decomposition (~15 L H$_2$/g$_{cat}$·h) was the same as in DMM, DME, methanol SR reactions. Taking into account these H$_2$ productivity data, we calculated that just ~50 g of the CuO-CeO$_2$/γ-Al$_2$O$_3$ catalytic system is enough to operate a 1 kW HT PEM FC-based power unit using any substrate—DMM + H$_2$O, DME + H$_2$O, methanol + H$_2$O or FA.

Table 3. Performance of the CuO-CeO$_2$/γ-Al$_2$O$_3$ catalyst in dimethoxymethane, dimethyl ether, methanol steam reforming and FA decomposition.

| Reactions | Inlet Composition | T | GHSV | Y (H$_2$) | CO | W (H$_2$) | Refs |
	vol.%	°C	h^{-1}	%	vol.%	l/(h g$_{cat}$)	
DMM SR	C$_3$H$_8$O$_2$:H$_2$O:N$_2$ = 14:70:16	300	10,000	95	0.5	15.5	[12]
DME SR	C$_2$H$_6$O:H$_2$O:N$_2$ = 20:60:20	370		90	1	15	
Methanol SR	CH$_3$OH:H$_2$O:N$_2$ = 40:40:20	300		95	1	15	
FA decomposition	HCOOH:N$_2$ = 25:75	300	35,000	98	0.45	15	This work

4. Conclusions

A CuO-CeO$_2$/γ-Al$_2$O$_3$ catalyst for vapor-phase formic acid decomposition to H$_2$-rich gas is suggested. The catalyst containing CuO-CeO$_2$ mixed oxides on the γ-Al$_2$O$_3$ surface is active and selective in formic acid decomposition to H$_2$-rich gas with low CO concentration (<0.5 vol.%). In particular, the CuO-CeO$_2$/γ-Al$_2$O$_3$ catalyst provides 100% conversion of formic acid with H$_2$ yield ~98% at 200–300 °C and GHSV = 3500–350,000 h^{-1}. Moreover, the CuO-CeO$_2$/γ-Al$_2$O$_3$ catalyst shows high prospects for the multifuel processor concept. It enables the H$_2$ generation from oxygenated compounds of C1 chemistry (dimethyl ether, methanol, dimethoxymethane and formic acid) under similar reaction conditions.

Author Contributions: S.B. suggested the main idea of the article; A.P. carried out experiments, exported the results and prepared an initial version; V.B. suggested the methodology of the article; V.S. is in charge of supervision, writing—review and editing of the article. All authors discussed the results and revised and corrected this article.

Funding: This work was supported by the Ministry of Science and Higher Education of the Russian Federation (project № AAAA-A17-117041710088-0).

Conflicts of Interest: The authors declare no conflict of interest.

References

1. Chandan, A.; Hattenberger, M.; El-kharouf, A.; Du, S.; Dhir, A.; Self, V.; Pollet, B.G.; Ingram, A.; Bujalski, W. High temperature (HT) polymer electrolyte membrane fuel cells (PEMFC)—A review. *J. Power Sources* **2013**, *231*, 264–278. [CrossRef]

2. Quartarone, E.; Mustarelli, P. Polymer fuel cells based on polybenzimidazole/H_3PO_4. *Energy Environ. Sci.* **2012**, *5*, 6436–6444. [CrossRef]

3. Rosli, R.E.; Sulong, A.B.; Daud, W.R.W.; Zulkifley, M.A.; Husaini, T.; Rosli, M.I.; Majlan, E.H.; Haque, M.A. A review of high-temperature proton exchange membrane fuel cell (HT-PEMFC) system. *Int. J. Hydrogen Energy* **2017**, *42*, 9293–9314. [CrossRef]

4. Li, D.; Li, X.; Gong, J. Catalytic Reforming of Oxygenates: State of the Art and Future Prospects. *Chem. Rev.* **2016**, *116*, 11529–11653. [CrossRef] [PubMed]

5. Palo, D.R.; Dagle, R.A.; Holladay, J.D. Methanol Steam Reforming for Hydrogen Production. *Chem. Rev.* **2007**, *107*, 3992–4021. [CrossRef]

6. Sá, S.; Silva, H.; Brandão, L.; Sousa, J.M.; Mendes, A. Catalysts for methanol steam reforming—A review. *Appl. Catal. B.* **2010**, *99*, 43–57. [CrossRef]

7. Galvita, V.; Semin, G.; Belyaev, V.; Yurieva, T.; Sobyanin, V. Production of hydrogen from dimethyl ether. *Appl. Catal. A.* **2001**, *216*, 85–90. [CrossRef]

8. Semelsberger, T.A.; Ott, K.C.; Borup, R.L.; Greene, H.L. Generating hydrogen-rich fuel-cell feeds from dimethyl ether (DME) using Cu/Zn supported on various solid-acid substrates. *Appl. Catal. A.* **2006**, *309*, 210–223. [CrossRef]

9. Volkova, G.; Badmaev, S.; Belyaev, V.; Plyasova, L.; Budneva, A.; Paukshtis, E.; Zaikovsky, V.; Sobyanin, V. Bifunctional catalysts for hydrogen production from dimethyl ether. *Stud. Surf. Sci. Catal.* **2007**, *167*, 445–450.

10. Sun, Q.; Auroux, A.; Shen, J. Surface acidity of niobium phosphate and steam reforming of dimethoxymethane over $CuZnO/Al_2O_3$–NbP complex catalysts. *J. Catal.* **2006**, *244*, 1–9. [CrossRef]

11. Fu, Y.; Shen, J. Production of hydrogen by catalytic reforming of dimethoxymethane over bifunctional catalysts. *J. Catal.* **2007**, *248*, 101–110. [CrossRef]

12. Badmaev, S.D.; Pechenkin, A.A.; Belyaev, V.D.; Ven'yaminov, S.A.; Snytnikov, P.V.; Sobyanin, V.A.; Parmon, V.N. Steam reforming of dimethoxymethane to hydrogen-rich gas for fuel cell feeding application. *Dokl. Phys. Chem.* **2013**, *452*, 251–253. [CrossRef]

13. Pechenkin, A.A.; Badmaev, S.D.; Belyaev, V.D.; Sobyanin, V.A. Performance of bifunctional $CuO–CeO_2/\gamma\text{-}Al_2O_3$ catalyst in dimethoxymethane steam reforming to hydrogen-rich gas for fuel cell feeding. *Appl. Catal. B.* **2015**, *166*, 535–543. [CrossRef]

14. Badmaev, S.D.; Pechenkin, A.A.; Belyaev, V.D.; Sobyanin, V.A. Hydrogen production by steam reforming of dimethoxymethane over bifunctional $CuO\text{-}ZnO/\gamma\text{-}Al_2O_3$ catalyst. *Int. J. Hydrogen Energy* **2015**, *40*, 14052–14057. [CrossRef]

15. Joó, F. Breakthroughs in Hydrogen Storage-Formic Acid as a Sustainable Storage Material for Hydrogen. *ChemSusChem* **2008**, *1*, 805–808. [CrossRef] [PubMed]

16. Bulushev, D.A.; Beloshapkin, S.; Ross, J.R.H. Hydrogen from formic acid decomposition over Pd and Au catalysts. *Catal. Today* **2010**, *154*, 7–12. [CrossRef]

17. Zhou, X.; Huang, Y.; Xing, W.; Liu, C.; Liao, J.; Lu, T. High-quality hydrogen from the catalyzed decomposition of formic acid by Pd–Au/C and Pd–Ag/C. *Chem. Commun.* **2008**, *30*, 3540–3542. [CrossRef]

18. Tedsree, K.; Li, T.; Jones, S.; Chan, C.W.A.; Yu, K.M.K.; Bagot, P.A.J.; Marquis, E.A.; Smith, G.D.W.; Tsang, S.C.E. Hydrogen production from formic acid decomposition at room temperature using a Ag–Pd core–shell nanocatalyst. *Nat. Nanotechnol.* **2011**, *6*, 302–307. [CrossRef]

19. Fellay, C.; Dyson, P.; Laurenczy, G. A Viable Hydrogen-Storage System Based On Selective Formic Acid Decomposition with a Ruthenium Catalyst. *Angew. Chem. Int. Ed.* **2008**, *47*, 3966–3968. [CrossRef]

20. Bulushev, D.A.; Chuvilin, A.L.; Sobolev, V.I.; Stolyarova, S.G.; Shubin, Y.V.; Asanov, I.P.; Ishchenko, A.V.; Magnani, G.; Riccò, M.; Okotrub, A.V.; et al. Copper on carbon materials: Stabilization by nitrogen doping. *J. Mater. Chem. A.* **2017**, *5*, 10574–10583. [CrossRef]

21. Gazsi, A.; Bánsági, T.; Solymosi, F. Decomposition and Reforming of Formic Acid on Supported Au Catalysts: Production of CO-Free H_2. *J. Phys. Chem. C.* **2011**, *115*, 15459–15466. [CrossRef]

22. Solymosi, F.; Koós, Á.; Liliom, N.; Ugrai, I. Production of CO-free H_2 from formic acid. A comparative study of the catalytic behavior of Pt metals on a carbon support. *J. Catal.* **2011**, *279*, 213–219. [CrossRef]

23. Iglesia, E.; Boudart, M. Decomposition of formic acid on copper, nickel, and copper-nickel alloys: II. Catalytic and temperature-programmed decomposition of formic acid on Cu/SiO_2, Cu/Al_2O_3, and Cu powder. *J. Catal.* **1983**, *81*, 214–223. [CrossRef]

24. Caputo, T.; Lisi, L.; Pirone, R.; Russo, G. On the role of redox properties of CuO/CeO_2 catalysts in the preferential oxidation of CO in H_2-rich gases. *Appl. Catal. A. Gen.* **2008**, *348*, 42–53. [CrossRef]

25. Jiang, X.Y.; Lou, L.P.; Chen, Y.X.; Zheng, X.M. Effects of CuO/CeO_2 and CuO/γ-Al_2O_3 catalysts on NO + CO reaction. *J. Mol. Catal. A. Chem.* **2003**, *197*, 193–205.

26. Menon, U.; Poelman, H.; Bliznuk, V.; Galvita, V.V.; Poelman, D.; Marin, G.B. Nature of the active sites for the total oxidation of toluene by CuO-CeO_2/Al_2O_3. *J. Catal.* **2012**, *295*, 91–103. [CrossRef]

Article

The Role of Support in Formic Acid Decomposition on Gold Catalysts

Vladimir Sobolev [1,*], Igor Asanov [2,3] and Konstantin Koltunov [1,3]

[1] Boreskov Institute of Catalysis SB RAS, 630090 Novosibirsk, Russia; koltunov@catalysis.ru
[2] Nikolaev Institute of Inorganic Chemistry SB RAS, 630090 Novosibirsk, Russia; asan@niic.nsc.ru
[3] Department of Natural Sciences, Novosibirsk State University, 630090 Novosibirsk, Russia
* Correspondence: visobo@catalysis.ru; Tel.: +73-833-269-765

Received: 5 October 2019; Accepted: 1 November 2019; Published: 4 November 2019

Abstract: Formic acid (FA) can easily be decomposed, affording molecular hydrogen through a controllable catalytic process, thus attaining great importance as a convenient hydrogen carrier for hydrogen energetics. Supported gold nanoparticles are considered to be among the most promising catalysts for such applications. However, questions remain regarding the influence of the catalyst support on the reaction selectivity. In this study, we have examined the catalytic activity of typical gold catalysts, such as Au/TiO_2, Au/SiO_2, and Au/Al_2O_3 in decomposition of FA, and then compared it with the catalytic activity of corresponding supports. The performance of each catalyst and support was evaluated using a gas-flow packed-bed reactor. It is shown that the target reaction, $FA \rightarrow H_2 + CO_2$, is provided by the presence of gold nanoparticles, whereas the concurrent, undesirable pathway, such as $FA \rightarrow H_2O + CO$, results exclusively from the acid-base behavior of supports.

Keywords: hydrogen energetics; hydrogen carrier; formic acid dehydrogenation; supported gold catalysts

---

1. Introduction

Formic acid (FA) is one of the liquid organic hydrogen carriers (LOHC), which contains 4.4 wt% of hydrogen. This chemical is persistent and thus, can be safely stored, transported, and applied as a source of hydrogen for the needs of hydrogen energetics instead of difficult to handle molecular hydrogen [1–3]. Sufficient stability combined with low toxicity and low flammability provides additional advantages for the practical use of FA. It is important that FA can be produced sustainably from biomass [4,5] or via reaction of CO_2 with hydrogen, which can be obtained by electrolysis [6,7]. Another key point is that FA undergoes facile dehydrogenation over supported noble metal catalysts at mild conditions. In particular, it was reported that the activity of Au/Al_2O_3 catalysts was higher compared to that of Pt/Al_2O_3 catalysts in the gas-phase reaction [8]. Overall, the catalytic behavior of supported Au in the FA decomposition is influenced by the nature of the support [9,10], the gold dispersion [8,11], and the use of basic cocatalysts [12–14].

In principle, FA can be decomposed through the desired path (1) to afford a mixture of molecular hydrogen and carbon dioxide or alternatively, FA can undergo dehydration to give concurrently highly undesirable carbon monoxide (path 2):

$$HCOOH \rightarrow CO_2 + H_2 \tag{1}$$

$$HCOOH \rightarrow CO + H_2O \tag{2}$$

It is noteworthy that despite significant efforts in finding efficient homo- and heterogeneous catalysts to address pathway (1), the achievement of near-theoretical-maximum selectivity of this reaction is still a challenging task. On the other hand, it was demonstrated recently that gold catalysts with the same gold content (~2.5 wt%) and mean Au-particles size (2.4–3.0 nm) supported on MgO, La_2O_3, ZrO_2, CeO_2, and Al_2O_3 have exhibited a clear volcano-type dependence of the FA decomposition along path (1) on the electronegativity of the support's cation, with the Au/Al_2O_3 on the top [9]. Moreover, the latter catalyst has demonstrated the best selectivity and practically CO-free hydrogen production. Based on these findings, it was suggested that dehydrogenation of FA should be catalyzed by gold nanoparticles but only in combination with the acid-base activity of the support. It is significant in this context that the acid-base reactivity of metal oxides themselves toward dehydration/dehydrogenation of FA is also generally known [15–17]. However, the relative contribution of oxide support as part of the gold catalyst to pathways (1) and (2) remained uncertain.

In order to fill that gap, in this study, we have examined the catalytic behavior of gold catalysts, such as Au/Al_2O_3, Au/TiO_2, and Au/SiO_2 in decomposition of FA, and then compared it with the catalytic activity of corresponding supports. It is found that the undesirable pathway (2) results exclusively from the acid-base behavior of supports, regardless of the presence of gold.

2. Materials and Methods

Commercially available Al_2O_3 (A-201 La Roche Industries Inc.), SiO_2 (Merck), and TiO_2 (Aerolyst 7708 Degussa AG) have been used as oxide supports. The deposition of gold on the oxide supports was implemented by a direct ionic exchange method using ammonia according to a procedure elaborated by Ivanova et al. [18]. An aqueous 5×10^{-4} M solution of $HAuCl_4$ (99.9%, ABCR, Darmstadt, Germany) was mixed with a support in a ratio corresponding to the Au concentration of 2 wt%. After stirring at 70 °C for 2 h, a 4 M solution of ammonia was added. The suspension thus obtained was stirred at 70 °C for an additional 1 h, then filtered off and washed with water. The samples were dried overnight at 80 °C and calcined at 300 °C for 4 h.

The specific surface area of the catalysts was determined from N_2 adsorption isotherms at 77.3 K on a Quadrosorb evo (Quantachrome Instruments, USA) analyzer after degassing the samples with the use of a FLOVAC Degasser (Quantachrome Instruments, USA) instrument.

XPS study was performed with a SPECS Phoibos 150 (Germany) photoelectron spectrometer. The spectra were recorded using an $AlK\alpha$ source with quantum energy of 1486.6 eV. The energy scale was calibrated against the binding energy of gold $Au4f_{7/2}$ equal to 84.0 eV. A carbon C 1s peak for hydrocarbon impurities at 285.0 eV was used as a reference for the energy scale calibration.

The gold content in the catalysts was determined by inductively coupled plasma optical emission spectrometry (ICP-OES) using an Optima 5300 DV (Perkin-Elmer) instrument. The samples (50 mg) were dissolved in 5 ml of imperial water ($2HNO_3$ and 6HCl). The solutions obtained were diluted with water to 100 ml and then analyzed.

Transmission electron microscopy (TEM) was performed with a Zeiss LEO 912 OMEGA (Germany) microscope at an acceleration voltage of 120 kV. To determine the mean diameter of Au particles, more than 100 particles were chosen.

The gas-phase decomposition of formic acid was performed in a quartz tube flow reactor with an internal diameter of 6 mm. Activity tests were carried out at atmospheric pressure using 20 mg of a catalyst mixed with 0.5 mL of quartz sand (d = 0.25 mm). The overall length of the catalyst layer was ~20 mm. The gas mixture (5 vol% formic acid in He) was fed at a rate of 20 mL min^{-1}. The experiments were carried out in the temperature-programmed mode. The heating rate was 2 °C min^{-1}. The reaction temperature was measured with a thermocouple inserted into the catalyst bed. In order to achieve reproducible results, prior to activity measurements, the catalysts were pretreated in the reactor according to a standard procedure: heating at 300 °C in a flow of 5 vol% HCOOH in He for 15 min, which ensured the in situ reduction conditions. Then the catalysts were cooled in the same mixture to the reaction temperature. The progress of the reaction was monitored by gas chromatography, judged by the production of CO and CO_2, as described earlier [19].

3. Results

3.1. Catalysts Characterization

It is known that the most active gold catalysts contain small gold particles, which are less than 10 nm in diameter, especially on such supports as reducible TiO_2. With this in mind, samples of Au/TiO_2, Au/Al_2O_3, and Au/SiO_2 with a gold content of ~2 wt% and an average gold particle size below 3 nm were prepared. The gold particle size was estimated by TEM. The particle size distribution of Au/TiO_2, Au/Al_2O_3, and Au/SiO_2 samples are shown in Figure 1. The gold loading in the prepared samples was determined by using ICP-OES. The exact loadings of Au, the surface areas of the catalysts found by BET, and the average sizes of gold particles are stated in Table 1.

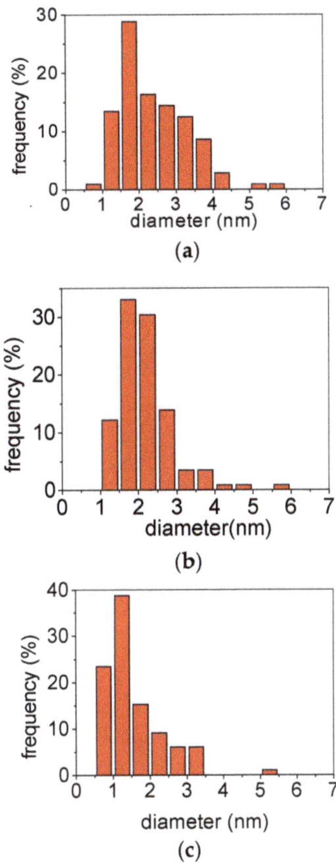

Figure 1. Particle size distribution of (**a**) Au/TiO_2, (**b**) Au/Al_2O_3, (**c**) Au/SiO_2 catalysts.

Table 1. Characteristics of Au-based catalysts and corresponding supports.

Sample	Au Loading (wt%)	S_{BET} (m$^2\cdot$g^{-1})	Average Size of Au Particles (nm)
TiO_2	-	45	-
Au/TiO_2	1.9	45	2.4 ± 0.9
Al_2O_3	-	200	-
Au/Al_2O_3	1.8	199	2.2 ± 1.0
SiO_2	-	480	-
Au/SiO_2	2.1	440	1.6 ± 0.8

The Au catalysts were also characterized by XPS in order to gain insight into the electronic state of Au. The XPS showed the presence of Au in the metallic state only (Au $4f_{7/2}$~84.0 eV) (Figure 2), while the nature of the support did not influence notably the electronic state of Au. However, the deconvolution of the XPS spectra revealed the existence of metallic gold with peaks shifted towards lower binding energies. This can be related to divers phenomena—the electron transfer from the support to particles of gold [20,21] and/or to a dominant effect of the atoms at low coordinated sites present in small particles, as it was proposed by Radnik et al. [22,23].

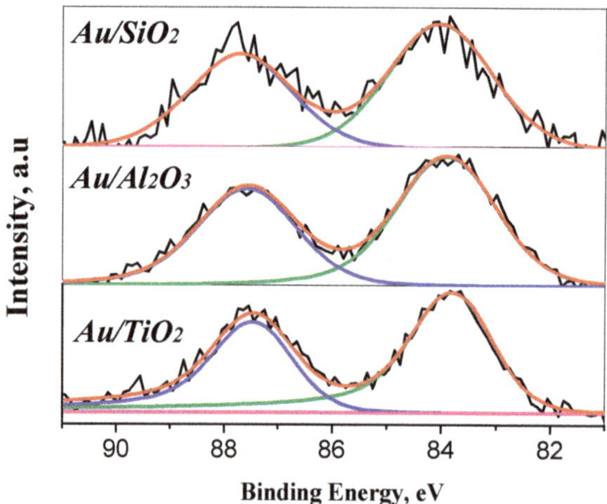

Figure 2. XPS of gold-containing catalysts in the Au 4f region.

3.2. Catalytic Activity

The catalytic performance of the oxide supported gold catalysts is shown in Figure 3. Despite the comparable content of gold (~2 wt%) and mean Au-particles size (~2 nm), the catalytic properties of Au/TiO_2, Au/SiO_2, and Au/Al_2O_3 samples in the FA decomposition are notably different. The relative activity of the catalysts turned out to be as follows: Au/TiO_2 > Au/SiO_2 > Au/Al_2O_3. Complete conversion of FA was achieved at 190 °C for Au/TiO_2, whereas for Au/Al_2O_3 and Au/SiO_2, the FA conversion had attained only 5 and 15% at the same conditions. However, decomposition of FA occurred by different pathways. The most active catalyst, Au/TiO_2, provided mainly undesirable pathway (2) to give CO and H_2O. For this catalyst, the selectivity towards H_2 formation was below 5% at complete conversion of FA. In contrast, decomposition of FA on Au/Al_2O_3 and Au/SiO_2 afforded mostly H_2 and CO_2 with a selectivity of ~80% (at complete conversion of FA).

For illustration, the long-term performance of Au/SiO_2 in the FA decomposition is presented in Figure 4. As can be seen, the catalytic activity and reaction selectivity practically did not change, even after FA dehydrogenation for 5 h at 228 °C.

Figure 5 shows the catalytic activity of Au/SiO_2, Au/Al_2O_3, and Au/TiO_2 samples versus that of the corresponding oxide supports. It is definitely seen that the catalytic performance of Au/TiO_2 relies almost entirely on the action of TiO_2 (Figure 5c). Unfortunately, TiO_2 (and therefore, Au/TiO_2) is more inclined to catalyze dehydration of FA on its acid sites [17], thus demonstrating comparatively poor catalytic activity towards the desirable path (1).

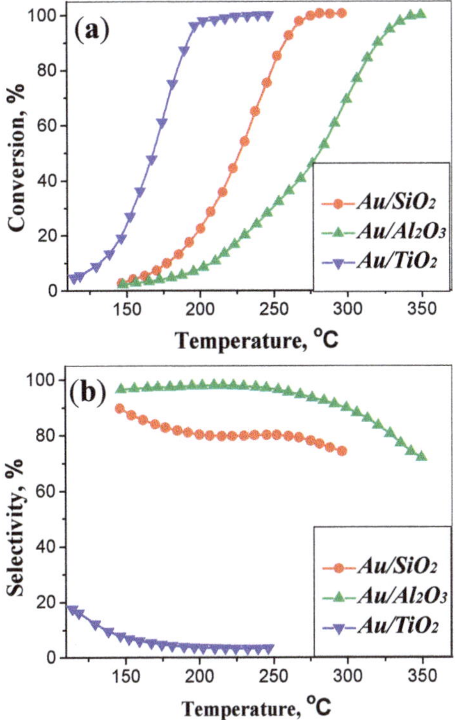

Figure 3. Conversion of formic acid (FA) (**a**) and selectivity to H_2 and CO_2 formation (**b**) in the decomposition of FA as a function of temperature on gold-containing catalysts.

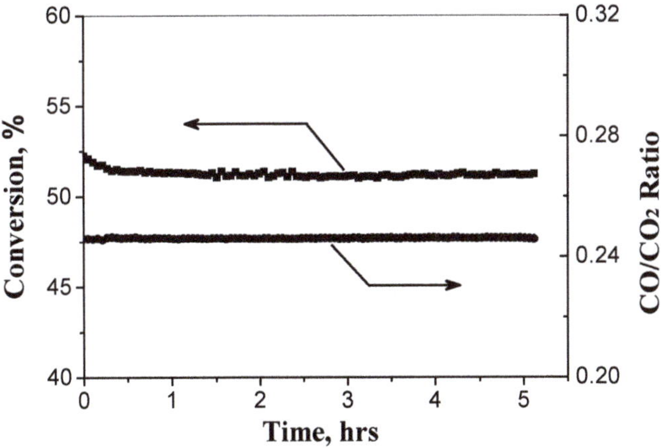

Figure 4. Long-term FA decomposition on Au/SiO_2 at 228 °C.

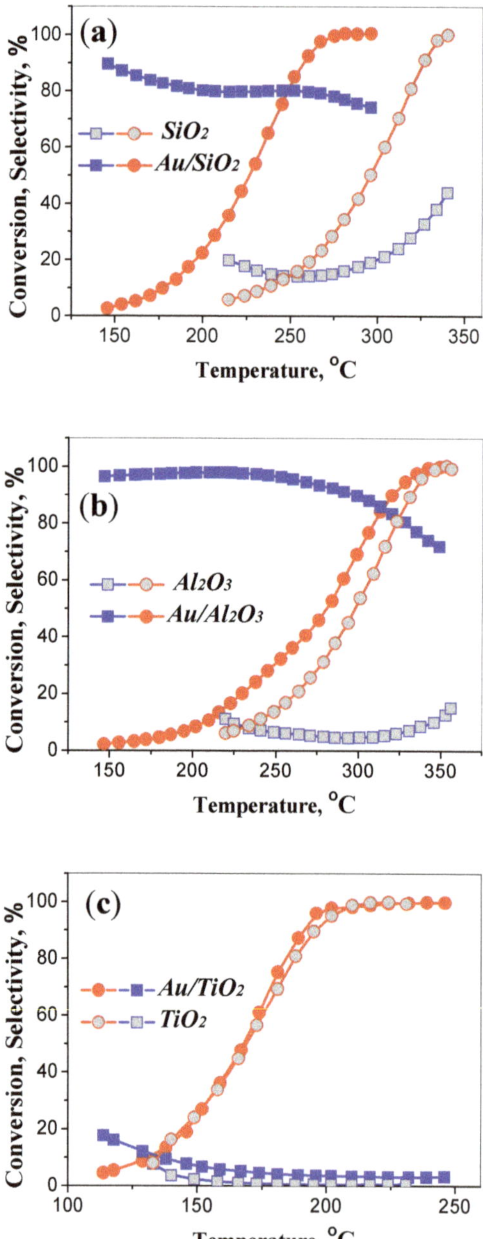

Figure 5. Conversions of FA (cycles) and selectivities to H_2 and CO_2 formation (squares) in the decomposition of FA on (**a**) Au/SiO$_2$ and SiO$_2$; (**b**) Au/Al$_2$O$_3$ and Al$_2$O$_3$; (**c**) Au/TiO$_2$ and TiO$_2$.

In contrast, SiO$_2$ itself has shown much lower catalytic activity compared to Au/SiO$_2$ (Figure 5a). The selectivity to FA dehydrogenation on SiO$_2$ is comparatively low and inversely associated with FA dehydration. Pure Al$_2$O$_3$ behaves in a similar way. The selectivity of FA conversion to H_2 and CO_2 is very low (Figure 5b). Thus, none of the oxide supports has demonstrated significant selectivity for H_2

production. On the contrary, their acid-base properties lead mainly to the undesirable decomposition of FA by pathway (2).

In general, these observations are in accordance with recent findings on the enhanced selectivity to hydrogen formation in FA decomposition over gold and other metals supported on an inert support, such as carbon or carbon doped with nitrogen [14,24–27]. However, these results were considered mainly in terms of the increased catalytic activity of corresponding catalysts toward FA dehydrogenation. In our vision, the oxide supports, such as SiO_2, Al_2O_3, and especially TiO_2, strongly compete with the target reaction, promoting acid-catalyzed dehydration of FA.

4. Conclusions

It is shown that the oxide support (regardless of whether it is TiO_2, Al_2O_3, or SiO_2) does not affect the electronic state of the supported nanoparticles of gold, which exist in the metallic state. However, a notable effect of the oxide support on the selectivity of hydrogen production from FA decomposition is found. The target reaction (FA $\rightarrow$ H_2 + CO_2) competes strongly with dehydration of FA due to the acid-base properties of the supports in the order given: $TiO_2 > Al_2O_3 \sim SiO_2$. Hence, the results of this study can be applied for the development of supported gold catalysts with the diminished acidity of supports for the selective hydrogen production via FA decomposition.

Author Contributions: Catalytic measurements, data analysis, and text editing, V.S.; data analysis and writing, K.K.; XPS measurements, I.A.

Funding: This work was conducted within the framework of the budget project for the Boreskov Institute of Catalysis (No. AAAA-A17-117041710083-5).

Conflicts of Interest: The authors declare no conflict of interest.

References

1. Gianotti, E.; Taillades-Jacquin, M.; Roziere, J.; Jones, D.J. High-Purity Hydrogen Generation Via Dehydrogenation of Organic Carriers: A Review on the Catalytic Process. *ACS Catal.* **2018**, *8*, 4660–4680. [CrossRef]

2. Niermann, M.; Beckendorff, A.; Kaltschmitt, M.; Bonhoff, K. Liquid Organic Hydrogen Carrier (LOHC)—Assessment Based on Chemical and Economic Properties. *Int. J. Hydrogen Energy* **2019**, *44*, 6631–6654. [CrossRef]

3. Zhong, H.; Iguchi, M.; Chatterjee, M.; Himeda, Y.; Xu, Q.; Kawanami, H. Formic Acid-Based Liquid Organic Hydrogen Carrier System with Heterogeneous Catalysts. *Adv. Sustain. Syst.* **2018**, *2*, 1700161. [CrossRef]

4. Bulushev, D.A.; Ross, J.R.H. Towards Sustainable Production of Formic Acid. *ChemSusChem* **2018**, *11*, 821–836. [CrossRef] [PubMed]

5. Preuster, P.; Albert, J. Biogenic Formic Acid as a Green Hydrogen Carrier. *Energy Technol.* **2018**, *6*, 501–509. [CrossRef]

6. Bulushev, D.A.; Ross, J.R.H. Heterogeneous Catalysts for Hydrogenation of CO_2 and Bicarbonates to Formic Acid and Formates. *Catal. Rev.* **2018**, *60*, 566–593. [CrossRef]

7. Wang, W.H.; Himeda, Y.; Muckerman, J.T.; Manbeck, G.F.; Fujita, E. CO_2 Hydrogenation to Formate and Methanol as an Alternative to Photo- and Electrochemical CO_2 Reduction. *Chem. Rev.* **2015**, *115*, 12936–12973. [CrossRef]

8. Ojeda, M.; Iglesia, E. Formic Acid Dehydrogenation on Au-Based Catalysts at near-Ambient Temperatures. *Angew. Chem. Int. Ed.* **2009**, *48*, 4800–4803. [CrossRef]

9. Zacharska, M.; Chuvilin, A.L.; Kriventsov, V.V.; Beloshapkin, S.; Estrada, M.; Simakov, A.; Bulushev, D.A. Support Effect for Nanosized Au Catalysts in Hydrogen Production from Formic Acid Decomposition. *Catal. Sci. Technol.* **2016**, *6*, 6853–6860. [CrossRef]

10. Gazsi, A.; Bansagi, T.; Solymosi, F. Decomposition and Reforming of Formic Acid on Supported Au Catalysts: Production of CO-Free H2. *J. Phys. Chem. C* **2011**, *115*, 15459–15466. [CrossRef]

11. Singh, S.; Li, S.; Carrasquillo-Flores, R.; Alba-Rubio, A.C.; Dumesic, J.A.; Mavrikakis, M. Formic Acid Decomposition on Au Catalysts: DFT, Microkinetic Modeling, and Reaction Kinetics Experiments. *AIChE J.* **2014**, *60*, 1303–1319. [CrossRef]

12. Bulushev, D.A.; Zacharska, M.; Guo, Y.; Beloshapkin, S.; Simakov, A. CO-Free Hydrogen Production from Decomposition of Formic Acid over Au/Al$_2$O$_3$ Catalysts Doped with Potassium Ions. *Catal. Commun.* **2017**, *92*, 86–89. [CrossRef]

13. Jia, L.; Bulushev, D.A.; Beloshapkin, S.; Ross, J.R.H. Hydrogen Production from Formic Acid Vapour over a Pd/C Catalyst Promoted by Potassium Salts: Evidence for Participation of Buffer-Like Solution in the Pores of the Catalyst. *Appl. Catal. B Environ.* **2014**, *160*, 35–43. [CrossRef]

14. Bulushev, D.A.; Sobolev, V.I.; Pirutko, L.V.; Starostina, A.V.; Asanov, I.P.; Modin, E.; Chuvilin, A.L.; Gupta, N.; Okotrub, A.V.; Bulusheva, L.G. Hydrogen Production from Formic Acid over Au Catalysts Supported on Carbon: Comparison with Au Catalysts Supported on SiO$_2$ and Al$_2$O$_3$. *Catalysts* **2019**, *9*, 376. [CrossRef]

15. Trillo, J.M.; Munera, G.; Criado, J.M. Catalytic decomposition of formic acid on metal oxides. *Catal. Rev.* **1972**, *7*, 51–86. [CrossRef]

16. Patermarakis, G. The parallel dehydrative and dehydrogenative catalytic action of γ-Al$_2$O$_3$ pure and doped by MgO. *Appl. Catal. A* **2003**, *252*, 231–241. [CrossRef]

17. Sobolev, V.I.; Koltunov, K.Y. Oxidative and non-oxidative degradation of C$_1$–C$_3$ carboxylic acids over V$_2$O$_5$/TiO$_2$ and MoVTeNb oxides: A comparative study. *Appl. Catal. A Gen.* **2013**, *466*, 45–50. [CrossRef]

18. Ivanova, S.; Pitchon, V.; Zimmermann, Y.; Petit, C. Preparation of Alumina Supported Gold Catalysts: Influence of Washing Procedures, Mechanism of Particles Size Growth. *Appl. Catal. A Gen.* **2006**, *298*, 57–64. [CrossRef]

19. Bulushev, D.A.; Chuvilin, A.L.; Sobolev, V.I.; Stolyarova, S.G.; Shubin, Y.V.; Asanov, I.P.; Ishchenko, A.V.; Magnani, G.; Ricco, M.; Okotrub, A.V.; et al. Copper on Carbon Materials: Stabilization by Nitrogen Doping. *J. Mater. Chem. A* **2017**, *5*, 10574–10583. [CrossRef]

20. Claus, P.; Bruckner, A.; Mohr, C.; Hofmeister, H. Supported Gold Nanoparticles from Quantum Dot to Mesoscopic Size Scale: Effect of Electronic and Structural Properties on Catalytic Hydrogenation of Conjugated Functional Groups. *J. Am. Chem. Soc.* **2000**, *122*, 11430–11439. [CrossRef]

21. Sanchez, A.; Abbet, S.; Heiz, U.; Schneider, W.-D.; Hakkinen, H.; Barnett, R.N.; Landman, U. When Gold Is Not Noble: Nanoscale Gold Catalysts. *J. Phys. Chem. A* **1999**, *103*, 9573–9578. [CrossRef]

22. Radnik, J.; Mohr, C.; Claus, P. On the Origin of Binding Energy Shifts of Core Levels of Supported Gold Nanoparticles and Dependence of Pretreatment and Material Synthesis. *Phys. Chem. Chem. Phys.* **2003**, *5*, 172–177. [CrossRef]

23. Pohl, M.-M.; Radnik, J.; Schneider, M.; Bentrup, U.; Linke, D.; Bruckner, A.; Ferguson, E. Bimetallic PdAu–KOAc/SiO$_2$ catalysts for vinyl acetate monomer (VAM) synthesis: Insights into deactivation under industrial conditions. *J. Catal.* **2009**, *262*, 314–323. [CrossRef]

24. Tang, C.; Surkus, A.-E.; Chen, F.; Pohl, M.-M.; Agostini, G.; Schneider, M.; Junge, H.; Beller, M. A Stable Nanocobalt Catalyst with Highly Dispersed CoN$_x$ Active Sites for the Selective Dehydrogenation of Formic Acid. *Angew. Chem. Int. Ed.* **2017**, *56*, 16616–16620. [CrossRef]

25. Sanchez, F.; Alotaibi, M.H.; Motta, D.; Chan-Thaw, C.E.; Rakotomahevitra, A.; Tabanelli, T.; Roldan, A.; Hammond, C.; He, Q.; Davies, T.; et al. production from formic acid decomposition in the liquid phase using Pd nanoparticles supported on CNFs with different surface properties. *Sustain. Energy Fuels* **2018**, *2*, 2705–2716. [CrossRef]

26. Sanchez, F.; Motta, D.; Roldan, A.; Hammond, C.; Villa, A.; Dimitratos, N. Hydrogen Generation from Additive-Free Formic Acid Decomposition Under Mild Conditions by Pd/C: Experimental and DFT Studies. *Top. Catal.* **2018**, *61*, 254–266. [CrossRef]

27. Sanchez, F.; Motta, D.; Bocelli, L.; Albonetti, S.; Roldan, A.; Hammond, C.; Villa, A.; Dimitratos, N. Investigation of the Catalytic Performance of Pd/CNFs for Hydrogen Evolution from Additive-Free Formic Acid Decomposition. *C J. Carbon Res.* **2018**, *4*, 26. [CrossRef]

Review

Hydrogen Production from Formic Acid Attained by Bimetallic Heterogeneous PdAg Catalytic Systems

Miriam Navlani-García [1,*], David Salinas-Torres [2] and Diego Cazorla-Amorós [1]

[1] Department of Inorganic Chemistry and Materials Institute, University of Alicante, 03080 Alicante, Spain; cazorla@ua.es

[2] Department of Physical Chemistry and Materials Institute, University of Alicante, 03080 Alicante, Spain; david.salinas@ua.es

* Correspondence: miriam.navlani@ua.es; Tel.: +34-965-903-400 (ext. 9150)

Received: 26 September 2019; Accepted: 21 October 2019; Published: 23 October 2019

Abstract: The production of H_2 from the so-called Liquid Organic Hydrogen Carriers (LOHC) has recently received great focus as an auspicious option to conventional hydrogen storage technologies. Among them, formic acid, the simplest carboxylic acid, has recently emerged as one of the most promising candidates. Catalysts based on Pd nanoparticles are the most fruitfully investigated, and, more specifically, excellent results have been achieved with bimetallic PdAg-based catalytic systems. The enhancement displayed by PdAg catalysts as compared to the monometallic counterpart is ascribed to several effects, such as the formation of electron-rich Pd species or the increased resistance against CO-poisoning. Aside from the features of the metal active phases, the properties of the selected support also play an important role in determining the final catalytic performance. Among them, the use of carbon materials has resulted in great interest by virtue of their outstanding properties and versatility. In the present review, some of the most representative investigations dealing with the design of high-performance PdAg bimetallic heterogeneous catalysts are summarised, paying attention to the impact of the features of the support in the final ability of the catalysts towards the production of H_2 from formic acid.

Keywords: hydrogen production; formic acid; heterogeneous catalysts; bimetallic nanoparticles; PdAg; AgPd; alloy

---

1. Introduction

If we think about the major problems faced by mankind, many socio-political matters, such as the economy, wars, terrorism, etc., would come to our minds. However, people are much less aware of environmental issues, which, for some people, are still a remote concept far from being within their concerns. Even if we do not "perceive" the global warming in our day-to-day existence, catastrophic consequences, such as the ozone depletion, or the extinction of species and destruction of marine habitat and deforestation, are a fact that jeopardises the continuity of our life as we know it nowadays. Never before our planet and existence have depended to such an extent on how we cope with climate change. Creating a sustainable world society is mandatory to reduce the human impact on nature, and using scientific knowledge as a guide is the only option. Many actions could be taken to reduce the anthropogenic impact on the environment. However, since mankind's energy production and, in particular, the combustion of fossil fuels (coal, oil, and natural gas), is the main responsible for the anthropogenic emission of greenhouse gases, the transition towards renewable and carbon-free energy sources is required in the present environmental context to minimise hazardous anthropogenic interference with the natural climate system. Even though such necessity is an obvious fact, it is challenging not only for developed countries in which the living standard requires large energy supplies, but also for those developing countries that experience a high population growth.

Among greenhouse gases, carbon dioxide (CO_2) is considered the most important because, even though its global warming potential (GWP) is much lower than that of other gases, the emissions of CO_2 are far more abundant. GWP parameter was developed to compare the global warming impacts of different gases and it is a measure of how much energy the emissions of 1 ton of a gas will absorb over a given period, relative to the emissions of 1 ton of CO_2, which is used as the reference. Then, CO_2 has a GWP of 1 regardless of the time used, while GWP is 28–36 and 265–298 times that of CO_2 for methane (CH_4) and nitrous oxide (N_2O), respectively [1]. Such considerations, that are nowadays central scientific and social concerns, have a long historical background within the research community. Arrhenius was among the first to hypothesise about the impact of CO_2 on the Earth's climate [2], but his hypothesis was overlooked until the 1950s [3]. Now, after more than half a century, there is not yet a chemical process able to efficiently clean the growing volumes of CO_2 in the atmosphere. Some interesting actions taken so far are related to the production of hydrocarbon fuels by recycling H_2O and CO_2 with renewable energy sources or the CO_2 capture and sequestration (CCS) technology [4–7].

In such energy and environmental context, the search for alternative carbon emission-free energy sources is required to avoid further disastrous consequences. Hydrogen (H_2), a carbon-free fuel, is considered an ideal energy vector. However, the job is far from done and there are still important limitations for the implementation of hydrogen into the global energy scenario. Moreover, H_2 is not readily available in nature and, even though it can be generated from renewable sources (i.e. gasification of biomass/biofuels and water splitting), it is mostly produced via steam reforming process, which indeed is not sustainable because it uses fossil fuels and CO_2 is produced [8]. Moreover, aside from that, there are also significant limitations related to its difficult storage and delivery [9], as well as safety issues derived from its high flammability and potential explosiveness [10]. A promising alternative to overcome those limitations is the production of H_2 from hydrogen carrier molecules. Among them, the so-called Liquid Organic Hydrogen Carriers (LOHC) have recently received great attention as an interesting option to conventional hydrogen technologies, with particular importance for mobile applications. In such molecules, H_2 is stored and delivered through reversible hydrogenation and dehydrogenation chemical reactions upon utilisation of suitable catalysts [11]. Among the virtues of LOHC is the fact that they may use the existing infrastructure for fuel [12]. Although the first studies on LOHCs took place in the early 1980s, a renewed interest has emerged in the last years [11,13]. Bessarabov et al. recently summarised the properties of a "good" LOHC as follows [12]: (1) Low melting point (< −30 °C) and high boiling point (> 300 °C); (2) High hydrogen storage capacity; (3) Ability to undergo very selective hydrogenation and dehydrogenation reactions for long life cycles of charging and discharging; (4) Toxicological and eco-toxicological safety; (5) Low production costs and good technical availability. Formic acid (methanolic acid, HCOOH, FA), the simplest carboxylic acid (pK_a = 3.75), is an outstanding LOHC candidate, which not only encompasses most of the features listed above but also, it is a product of the hydrogenation of CO_2 [14,15].

The relative moderate hydrogen capacity of 4.4 wt.% might seem to be its weak point as compared to other LOHC [16]. However, its high density of 1.22 g/cm^3 leads to a volumetric capacity of 53 gH_2/L, equivalent to an energy density of 1.77 kW·h/L, which is larger than the value of compressed hydrogen at 70 MPa [9,17].

The production of H_2 from FA was firstly investigated by Coffey in 1967 [18], but it was not until 2008 when the interest of FA as a LOHC was renewed thanks to the independent investigations carried out by Laurenczy and Beller [19–21]. Along these years, studies dealing with the production of H_2 from FA have been carried out by using both, homogeneous and heterogeneous catalysts of diverse composition [22–24]. Due to their significance within the catalytic systems applied to this reaction, this paper reviews some of the most representative studies dealing with the decomposition of FA using PdAg-based bimetallic catalysts.

2. Decomposition of Formic Acid

The decomposition of FA can follow two pathways according to the following reactions:

$$\text{Dehydrogenation (decarboxylation): } HCOOH \leftrightarrow H_2 + CO_2; \ \Delta G^\circ = -32.9 \text{ kJ mol}^{-1} \tag{1}$$

$$\text{Dehydration (decarbonylation): } HCOOH \leftrightarrow H_2O + CO; \ \Delta G^\circ = -12.4 \text{ kJ mol}^{-1} \tag{2}$$

The dehydrogenation path is the target reaction for the production of H_2, while dehydration should be avoided due to the poisonous effect of CO. The yield of each reaction could be affected by the reaction conditions (i.e. temperature, concentration, presence of additives, etc.) [25], as well as by the features of the catalytic system used [26].

The decomposition of FA has traditionally been tackled using homogeneous catalysts. Noteworthy in this field are the studies performed by Laurenczy and Beller [25,27,28]. However, their obvious limitations in terms of practical application and recovery capabilities have motivated the search for efficient and selective heterogeneous systems able to produce H_2 from FA under mild conditions. In this sense, there is an increasing number of publications reporting on the investigation of catalytic systems able to catalyse the reaction while exploring the features of either support or metal active phase. Special mention should be made of those breakthroughs achieved Xu et al. [29–34], Mori et al. [35–46] and Bulushev et al. [47–57].

Most of the heterogeneous catalysts used are based on metal nanoparticles immobilised on supports of diverse nature (i.e. carbon materials, MOF, zeolites, resins, etc.) [23,58,59]. Among the various compositions of the metal active phase analysed so far, Pd nanoparticles are unquestionably the most widely employed since they display both high conversion of FA and selectivity values at moderate temperatures. However, the main weakness of those Pd-based monometallic catalytic systems is the low stability due to quick deactivation by adsorption of reaction intermediates and CO. Such negative effect can be reduced by modification of the properties of Pd nanoparticles by incorporating a second element to form alloy-structures or core-shell Pd-based catalysts. Among those possible compositions, PdAg bimetallic nanoparticles should be highlighted due to their outstanding performance. Furthermore, it is well-known that the final behaviour of the catalysts will not be only given by the active metal phase, but will also depend on the support used, which opens up countless options for tailoring the catalytic properties of the final materials. Herein, some of the most representative approaches followed towards the development of high-performance heterogeneous PdAg catalysts for the decomposition of FA will be reviewed.

3. Decomposition of Formic Acid over PdAg Bimetallic Catalysts

Heterogeneous catalysts formed by bimetallic Pd-based nanoparticles are promising catalysts towards the decomposition of FA. In particular, those systems based on Pd in conjunction with other noble metal have been widely investigated. PdAg-based catalysts deserve the pole position among those bimetallic systems used in this application. They have been proved to be one of the most successful options for attaining high catalytic activity by virtue of the resulting electron-rich Pd species caused by electron transfer from Ag to Pd in the nanoparticles (the so-called "ligand effect"), which is driven by the net difference in electronegativity (2.2 and 1.9 in Pauling scale, for Pd and Ag, respectively). It has been demonstrated that such electron-rich Pd species promote the cleavage of the O-H bond of FA molecules and favour the formation of formate intermediates. Detailed studies on the reaction mechanisms by kinetic isotope effect (KIE) using HCOOH, HCOOD, and DCOOH [35,37] were very useful to relate the presence of such electron rich species with the improved catalytic performance resulting from the facilitated O-H and C-H bond dissociations. Furthermore, studies in which experimental results were combined with theoretical calculations were also useful to relate the presence of electron-rich Pd species of bimetallic nanoparticles with the enhancement observed in the final catalytic activity resulting from the favoured O-H and C-H cleavage [38]. Furthermore, lattice contraction of the Pd surface by incorporation of a second element can also modify the adsorption site preference of FA molecules and suppress the formation of CO [60,61]. Aside from the features of the resulting Pd species, an additional benefit of these systems is that Pd and Ag can dissolve in each other to form bimetallic alloys with a huge range of compositions [62], which allows the preparation of catalysts with well-defined composition. Promising results have also been achieved with PdAu bimetallic nanoparticles, but the higher cost of Au as compared to Ag hinders its use.

The investigation conducted by Tsang et al. was one of the first studies reporting on the benefits of PdAg catalysts [26]. Three adsorption modes of FA molecules on the surface of the nanoparticles, i.e. monodentate (linear), bidentate (bridging), and multidentate (multilinear), were identified in that study, which were claimed to be prevalent on different sites. According to what they observed, the bridging form was prevalent on terrace sites and it favoured the dehydrogenation path ($H_2 + CO_2$), while monodentate and multidentate modes were favoured on surface-unsaturated sites and boost the dehydration pathway ($H_2O + CO$). Based on that observation, Ag–Pd core-shell nanocatalysts were designed and applied. The electronically modified Pd species originated from the charge transfer from Ag cores were responsible for strengthening the adsorption of bridging modes and for the subsequent enhancement of the catalytic performance. Among other characterization techniques, that study included Fourier transform infrared (FTIR), which was very useful to relate the performance of the catalysts with their properties. The promising results achieved in that study boosted the search for new PdAg-based catalysts. Among the most representative studies are those using carbon materials as supports [58], but some other interesting materials have also been used. In the following sections, representative studies on the decomposition of FA over PdAg-based catalysts supported on diverse materials are addressed.

In order to provide the readers at a glance view of the results achieved in this field, Table 1 lists some of the most representative PdAg-based bimetallic heterogeneous catalysts studied so far for the dehydrogenation of FA, together with the experimental conditions used and TOF values obtained.

Table 1. PdAg heterogeneous catalytic systems for the dehydrogenation of FA.

Catalysts	Temperature(°C)	Additive	Concentration of FA and Additive [a]	TOF (h^{-1})	Reference
AgPd@MIL-100(Fe)	25	-	1 M	58	[63]
PdAg/C–FA	25	HCOONa	5 M + 5 M	90 [b]	[64]
$Ag_{0.1}$-$Pd_{0.90}$/rGO	25	HCOONa	1 M + 0.67 M	105 [b]	[65]
C-Pd_1Ag_1 BNSs	25	HCOONa	1 M + 0.5 M	156	[66]
Ag_6Pd_1/N-rGO	25	-	5 M	171	[67]
PdAg-MnO_x/N-SiO_2	25	-	0.25 M	330 [b]	[62]
AgPd-Hs/G	25	HCOONa	2.5 M + 2.5 M	333	[68]
Ag@Pd/N-GCNT	25	-	N/A	413	[69]
$Ag_{74}Pd_{26}$/graphene	25	HCOONa	0.9 M + 0.1 M	572	[70]
PdAg-MnOx/N-SiO_2	30	-	0.25 M	530 [b]	[62]
AgPd/MOF-5-C	30	HCOONa	1.25 M + 3.75 M	854	[71]
PdAg-CeO_2/MC	30	HCOONa	N/A	2272	[72]
PdAg-MnO_x/N-SiO_2	35	-	0.25 M	700 [b]	[62]
PdAg-MnO_x/N-SiO_2	40	-	0.25 M	1430 [b]	[62]
C-Pd_1Ag_1 BNSs	50	HCOONa	1 M + 0.5 M	378	[66]
C-$Ag_{42}Pd_{58}$	50	-	1 M	382 [b]	[73]
Ag_9Pd_{91}/g-C_3N_4	50	HCOONa	1.5 M + 0.5 M	480	[74]
$Pd_{50}Ag_{50}$/Fe_3O_4/N-rGO	50	-	1.33 M	497	[75]
$Ag_{0.25}$Pd/WO_3	50	HCOONa	0.15 M + 1.2 M	683	[76]
$Ag_{10}Pd_{90}$/0.2CND/SBA-15	50	HCOONa	1.5 M + 0.5 M	893	[77]
Ag_1Pd_9/SBA-15-Amine	50	HCOONa	1.5 M + 0.5 M	964	[78]
Ag_1Pd_9@NPC	50	HCOONa	N/A	3000	[79]
Ag_1Pd_9–MnO_x/carbonspheres	50	HCOOK	5 M + 15 M	3558	[80]
PdAg@ZrO_2/C	50	HCOOK	2.5 M + 10 M	9206	[81]
$Ag_{74}Pd_{26}$/graphene	60	HCOONa	0.9 M + 0.1 M	572	[70]
PdAg@ZrO_2/C/rGO	60	HCOONa	3 M + 7.5 M	4500	[32]
Ag_1Pd_2/CN	75	-	2 M	621 [b]	[82]
Pd1Ag2/C(1)	75	HCOONa	0.9 M + 0.1 M	855 [b]	[43]
PdAg/amine-MSC	75	HCOONa	0.9 M + 0.1 M	5638	[15]
$Ag_{18}Pd_{82}$@ZIF-8	80	HCOONa	N/A	580	[83]
$Ag_{20}Pd_{80}$@MIL-101	80	HCOONa	N/A	848	[84]
Ag_1Pd_4@NH_2–UiO-66	80	-	1.25 M	893 [b]	[85]
AgPd@NPC	80	HCOONa	0.25 M + 0.25 M	936	[86]

Notes: [a] Expressed as concentration of FA (M) + concentration of additive (M); [b] Initial TOF values.

3.1. Carbon-supported PdAg Catalysts

The success in attaining high-performance catalysts lies, to a large extent, in controlling their final features in terms of composition, morphology, and electronic properties, as well as in finding the properties-activity relationship. For that, diverse and numerous approaches for the control of the features of both metal active phase and catalytic support have been tackled. Among investigated catalytic supports, the role of carbon materials should be highlighted. Their thermal stability, surface resistance in both basic and acidic medium, tunable porosity and surface chemistry, etc., provide great options towards the design of high-performance catalysts while exploring numerous approaches.

Li et al. investigated the effect of the reducing agent (sodium borohydride (SB), formic acid (FA), ascorbic acid (AA), and hydrogen) utilised during the preparation of catalysts based on PdAg nanoparticles supported on carbon (PdAg/C–SB, PdAg/C–FA, and PdAg/C–AA, and PdAg/C–H) [64]. It was observed that, while the wet-chemical reduction method provided similar average nanoparticle size in all cases (~5 nm), larger nanoparticles were attained by using gas-chemical reduction (7.5 nm). Besides, XPS analysis was used to get information about the surface composition of the nanoparticles. It was found that surface Pd/Ag molar ratios followed the order PdAg/C–AA > PdAg/C–FA > PdAg/C–H > PdAg/C–SB (2.97, 2.48, 2.34, and 0.81, compared to the nominal Pd/Ag molar ratio of 1), indicating the importance of the reducing agent in controlling the surface composition of PdAg nanoparticles. Among them, the catalyst synthesised by FA-reduction was the most active (initial TOF of 90 h^{-1} at 25 °C), what was claimed to be due to the presence of electron-rich abundant Pd atoms on the surface of the nanoparticles as well as to their alloyed state. Our group also reported on the impact of the surface Pd/Ag ratio in the final catalytic performance [43]. In that study, we used a promising synthetic strategy based on the preparation of colloidal PVP-capped PdAg nanoparticles through the reduction by solvent method. Firstly, composition-controlled colloidal nanoparticles were prepared by using four different Pd/Ag molar ratios (i.e. 1/0.5, 1/1, 1/2, and 1/4) and three PVP/Pd molar ratios (1/1, 5/1, and 10/1) and they were subsequently loaded on activated carbon. The characterisation of the catalysts revealed that the Pd/Ag surface ratio (for a given fixed nominal Pd/Ag molar ratio in the nanoparticles) was dependent on the amount of PVP used in the synthesis. It was found that, under the conditions used in that synthesis, Pd/Ag ratio decreased with the amount of PVP, which was related to the relative interaction of Pd and Ag with the polymer molecules. In that study, the catalyst synthesised with a Pd/Ag ratio of 1/2 and PVP/Pd of 1/1 was the most active among those studied, achieving an initial TOF value as high as 855 h^{-1} at 75 °C.

The importance of using composition-controlled monodisperse AgPd nanoparticles in attaining good catalytic performance was also pointed out by Sun et al. [73]. In that study, AgPd nanoparticles with an average size of 2.2 ± 0.1 nm were synthesised by co-reduction of metal precursors in oleylamine (OAm), oleic acid (OA) and 1-octadecene (ODE), and they were subsequently loaded on Ketjen carbon for their evaluation as heterogeneous catalysts in the decomposition of FA. The following composition of the nanoparticles was assessed: Pd, $Ag_{25}Pd_{75}$, $Ag_{42}Pd_{58}$, $Ag_{52}Pd_{48}$, $Ag_{60}Pd_{40}$, $Ag_{80}Pd_{20}$, and Ag. Control of the nanoparticle size was achieved by means of modifying the content of OA used in the synthesis. High total metal contents were used for the synthesis of the supported catalysts (i.e. 17 wt. % AgPd, 19 wt. % Pd, and 18 wt. % Ag, for C-AgPd, C-Pd, and C-Ag catalysts, respectively). Except for the sample with larger Ag content, all bimetallic systems displayed enhanced performance with respect to the monometallic counterpart, which was particularly important in $C-Ag_{42}Pd_{58}$ (initial TOF of 382 h^{-1} at 50 °C). Such enhancement was attributed to the small size of the nanoparticle (2.2 ± 0.1 and 4.5 ± 0.2 nm, for AgPd and Pd nanoparticles, respectively) and the synergistic Ag-Pd effect in the alloy catalysts. An aspect to highlight of that study is that the reusability tests revealed that around 90% of the initial activity of the catalysts was preserved after four consecutive catalytic runs.

Cheng et al. also investigated the performance of monodisperse PdAg-based bimetallic catalysts for the decomposition of FA by following a similar synthetic strategy [70]. In that case, AgPd nanoparticles (average nanoparticle size of 3.0 ± 0.2 nm) were prepared by co-reduction of $AgNO_3$ and $Pd(acac)_2$ in OAm, OA and ODE, and using *tert*-butylamine borane ($BH_3 \cdot C_4H_{11}N$, TBB) as a reducing agent. The as-synthesised nanoparticles were subsequently loaded on graphene to obtain AgPd/graphene catalysts with various compositions of the nanoparticles ($Ag_{92}Pd_8$/graphene, $Ag_{86}Pd_{14}$/graphene, $Ag_{74}Pd_{26}$/graphene, $Ag_{51}Pd_{49}$/graphene, and $Ag_{41}Pd_{59}$/graphene). Monometallic counterpart catalysts were also synthesised as references. The formation of alloy nanoparticles was confirmed by high-resolution TEM (HRTEM) analysis and UV–vis spectra, in which the surface plasmon resonance (SPR) of Ag located at 427 nm decreased considerably upon alloying with Pd. Among those compositions assessed, $Ag_{74}Pd_{26}$/graphene displayed the highest TOF value (572 h^{-1}, at room temperature). In addition, a catalyst with the same composition of the nanoparticles and supported on Ketjen carbon was also evaluated and it showed inferior performance than $Ag_{74}Pd_{26}$/graphene, which indicated the positive role of graphene as a 2D support in promoting the mass transport and electron transfer. Furthermore, it was also observed that the reducing agent used (TBB) was crucial in attaining monodisperse AgPd nanoparticles. The use of different reducing agents (i.e. triethylamine borane (TEB) or methylamine borane (MeAB)) resulted in larger nanoparticles and the subsequent activity decay. It that case, as in most of the catalysts reported so far for the decomposition of FA, the reusability was shown to be the critical point to be enhanced.

Apart from the reducing agent used in the synthesis and the presence of capping agents, the selection of suitable catalytic support is a well-known option for attaining small and well-dispersed metal nanoparticles. For instance, Jiang et al. selected reduced graphene oxide (rGO) as excellent support and powerful dispersion agent of bimetallic AgPd nanoparticles prepared by co-reduction of $AgNO_3$ and Na_2PdCl_4 with $NaBH_4$ [65]. TEM analysis revealed that the supported $Ag_{0.1}Pd_{0.9}$

nanoparticles had an average size of 6 nm, while severe aggregation was found for the counterpart free-nanoparticles. It was claimed that oxygen functional groups present in GO (i.e. carboxylic, carbonyl, and hydroxyl) were pivotal for controlling the size of the nanoparticles. In addition, XPS analysis revealed the strong metal-support interaction existing in the final catalysts. As results of those features, $Ag_{0.1}Pd_{0.9}$/rGO generated a large volume of gas (H_2 and CO_2) and an initial TOF of 105.2 h^{-1} at 25 °C, which was much larger than that of the free nanoparticles counterpart. The role of the support was confirmed by evaluating the catalytic performance of a physical mixture of rGO and $Ag_{0.1}Pd_{0.9}$. The decayed activity observed in that case was indicative of the important synergistic effect between support and nanoparticles and its role in attaining well-dispersed nanoparticles with better ability for boosting the production of H_2 from FA.

In an attempt to explore the behaviour of new carbon materials, Wang et al. investigated the utilisation of metal-organic framework (MOF) derived porous carbon as support of AgPd catalysts [71]. In that case, MOF-5 was used as precursor and template for the synthesis of nanoporous carbon *via* direct carbonisation at different temperatures (i.e. 700, 800, 900, 1000 °C), and AgPd nanoparticles were subsequently deposited from $AgNO_3$ and H_2PdCl_4. Nanoparticles with fixed composition were supported on carbon obtained from the calcination of MOF-5-C at different temperatures (catalysts denoted as Ag_3Pd_{12}/MOF-5-C-700, Ag_3Pd_{12}/MOF-5-C-800, Ag_3Pd_{12}/MOF-5-C-1000). Furthermore, a carbon-supported catalyst using Vulcan XC-72 (Ag_3Pd_{12}/XC-72) and catalysts using MOF-5-C-900 with various Ag/Pd ratios, were also prepared for comparison purposes (Pd_{15}/MOF-5-C-900, Ag_6Pd_9/MOF-5-C-900, and Ag_{15}/MOF-5-C-900), being the total metal loading of 15 wt.% in all cases. Catalyst Ag_3Pd_{12}/MOF-5-C-900 displayed the best performance among the investigated catalysts, with an initial TOF value of 854 h^{-1} at 30 °C. The results of the catalytic tests together with the characterisation of the samples make the authors of that study claim that the improved performance displayed by Ag_3Pd_{12}/MOF-5-C-900 as compared to other samples was due to the composition of the nanoparticles in that sample (i.e. alloy PdAg and monometallic Pd^0 and Ag^0) and their small size and good dispersion on the support. In addition, the role of sodium formate (SF) in the final catalytic activity was also studied by using various FA/SF ratios in the reaction medium (i.e. 1/1, 1/2, 1/3, and 1/4). It was postulated that the presence of SF increased the likelihood of contact between metal active sites and formate ions, which would, therefore, accelerate the kinetics of the FA decomposition reaction. A similar conclusion had previously been extracted by Cai et al. [87], who demonstrated that hydrogen produced by direct formate hydrolysis ($HCOO^- + H_2O \rightarrow H_2 + HCO_3^-$) was negligible (while using only formate ion in absence of FA in the reaction mixture), while the presence of formate greatly promoted the dehydrogenation of FA in FA/SF mixtures. Such beneficial effect was claimed to be due to the induction of a favourable adsorption orientation of FA on the catalysts to favour the dehydrogenation pathway in presence of formate ions. Moreover, the role of formate as reaction intermediate was also mentioned.

As for the morphology of the metal active phase, most of the studies reports on the use of sphere-shaped or cubo-octahedral "solid" nanoparticles with a relative low ratio of metal surface atoms in the total nanostructure. Decreasing the nanoparticle size is a well-known strategy for increasing the surface atoms exposed in the nanoparticles, but such an approach does not always lead to enhanced catalytic performances [88]. In some cases, the higher reactivity of smaller nanoparticles favours the adsorption of reaction intermediates, which leads to space exclusion and block the active sites, ultimately resulting in poorer catalytic activities as compared to less reactive larger nanoparticles [88]. Some other investigations are focused on increasing the ratio of surface-active sites by constructing other nanostructures. For example, Chen et al. reported on the preparation of graphene-supported AgPd hollow nanoparticles [68]. A one-pot wet-chemical co-reduction of GO and metal precursors was used in that case using trisodium citrate dihydrate as the stabiliser and L-ascorbic acid (L-AA) as the reducing agent (see Figure 1). Such a synthetic approach afforded the development of sphere-shaped hollow nanoparticles with a thin wall of 5 nm and an average size of 18 nm, which were well-dispersed on the surface of the catalyst (catalysts denoted as AgPd-Hs/G). Pd/G and AgPd/C catalysts were also prepared for comparison purposes in that study. XPS analysis revealed the existence of Pd^0, PdO and Ag^0 in AgPd-Hs/G. It was also observed that AgPd nanoparticles acted as nanoscaled spacers by increasing the interlayer spacing between graphene sheets and avoiding graphene layers stacking. In addition, it was also demonstrated that the formation of AgPd alloy nanoparticles was assisted by GO and the use of additives such as trisodium citrate dihydrate since such structures were not formed upon utilisation of other supports or in absence of additives. Regarding the catalytic activity, it was observed that AgPd-Hs/G generated as much as twice of the volume of gas generated by AgPd/C, which further confirmed the suitability of graphene as catalytic support in the present application. Besides, the importance of the hollow structure, in which most of the atoms are located on the surface of the nanoparticles, was also evidenced by comparison with a AgPd/G counterpart catalyst.

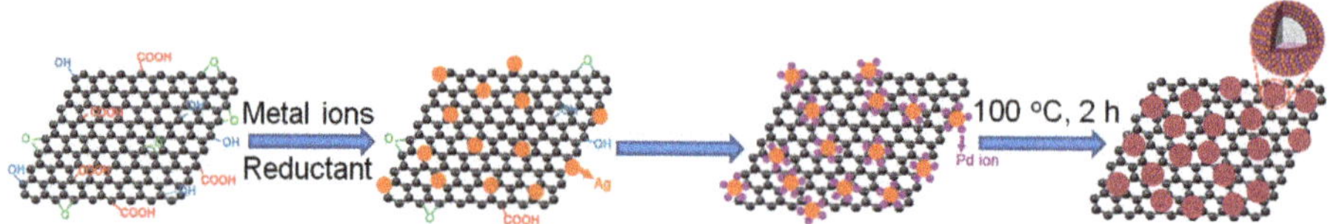

Figure 1. Schematic illustration for the preparation of AgPd-Hs/G. The preparation involves the redox reaction between the Pd ions and Ag nanoparticles. Reprinted with permission from [68].

Zheng et al. studied the synergetic effects of PdAg nanoparticles in the decomposition of FA by using two-dimensional ultrathin PdAg bimetallic nanosheets as model catalysts [66]. Such structures were selected to study the Pd-Ag synergic effect because, as in the case of hollow bimetallic nanostructures previously mentioned [68], they have a high fraction of active sites on the surface. Furthermore, the relation between the structure and the catalytic performance displayed by the nanosheets can be easily analysed due to the presence of a simple type of active sites. The preparation of PdAg bimetallic nanosheets (PdAg BNSs) was carried out in two steps.

Firstly, Pd NSs with hexagonal morphology were obtained from $Pd(acac)_2$, and after that, different contents of Ag (i.e. Pd/Ag ratios of 2/1, 1/1, 1/2, and 1/10) were chemically deposited on the as-prepared Pd NSs by using $AgNO_3$ and sodium citrate as metal precursor and reducing agent, respectively. The role of citrate in the synthesis was a key aspect for attaining alloy structures, due to its strong coordinating ability with Ag^+ and Pd^{2+}. The resulting PdAg BNSs preserved hexagonal morphology, but their thickness increased (as compared to Pd NSs) with the Ag content. After that, the obtained PdAg BNSs were loaded onto Vulcan XC-72 carbon to have a final Pd content of 5 wt.% in all cases (catalysts denoted as C-PdAg). In addition, monometallic Pd NSs (with a thickness of 1.8 nm and an average diameter of 85 nm) were also loaded on carbon (C-Pd) to check their activity in the reaction under evaluation. It was observed that C-Pd displayed a very poor activity and durability, which was attributed to the adsorption of CO produced via dehydration reaction on Pd (111) surface. The results of the catalytic performance achieved by C-PdAg catalysts is depicted in Figure 2. As can be seen in Figure 2a,b, the catalytic activity strongly depended on the composition of the catalysts, being optimum for C-PdAg with a ratio of 1/1. Further addition of Ag caused a marked activity decay due to excessive coverage of Pd active sites. The prominent catalytic activity of that sample was explained in terms of electronic promotion of Pd by Ag in the bimetallic structure. Besides, the activation energy of that catalyst was determined from experiments done at different temperatures (Figure 2c,d). Cyclic voltammograms (CV), CO stripping measurements and in-situ FTIR analysis helped to correlate the catalytic activity with the ability against CO poisoning.

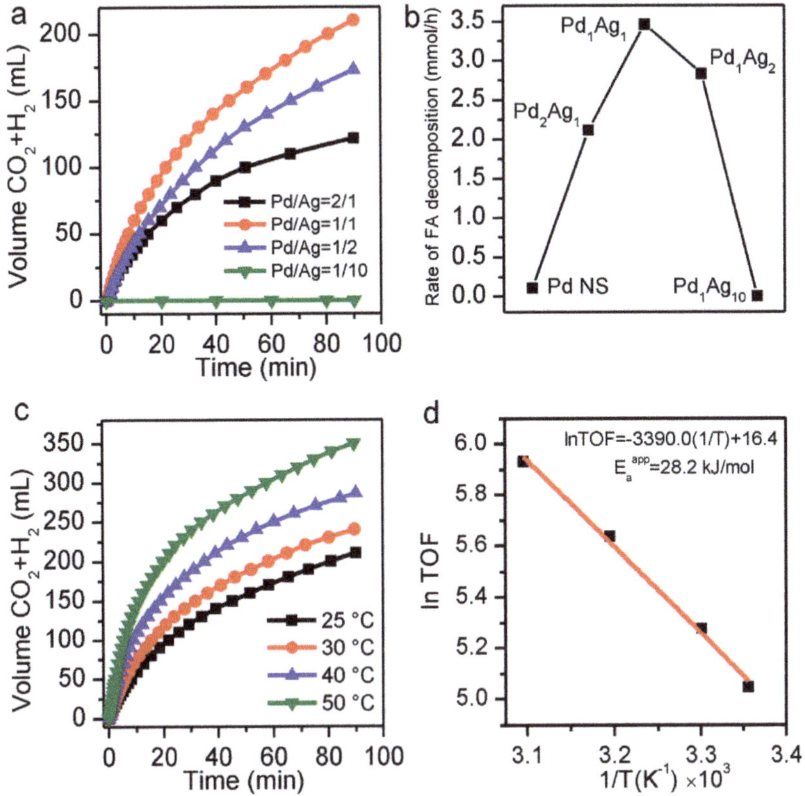

Figure 2. (**a**) Plots of the volume of generated gases (CO$_2$ +H$_2$) versus time from 10 mL solution of 1 M FA + 0.5 M SF in the presence of 50 mg C-PdAg BNSs with different Ag/Pd ratios (5 wt. % Pd for all catalysts) at 25 °C under ambient atmosphere. (**b**) Rate of FA decomposition catalysed by different C-PdAg BNSs. (**c**) Plots of the volume of generated gases (CO$_2$ + H$_2$) catalysed by C-Pd$_1$Ag$_1$ BNSs under different temperatures. (**d**) Arrhenius plot (ln (TOF) vs.1/T). Reprinted with permission from [66].

3.2. N-Doped Carbon-based PdAg Catalysts

The utilisation of "functional support" has widely been reported as an interesting option for achieving catalysts with tailored properties. As for the catalysts based on monometallic Pd nanoparticles, it was observed that the addition of nitrogen atoms into the carbon matrix promoted the electrical conductivity of the resulting material and serves as sites for the immobilisation of metal through strengthening the metal-support interaction. Furthermore, it has also been noted that, due to their basic character, nitrogen atoms can actively participate in the dehydrogenation of FA [16,53,57,89–91]. Then, due to the positive role of the addition of nitrogen functional groups in the support, it has also been a very fruitfully investigated strategy for the enhancement of the final performance of PdAg-based catalysts while considering supports of diverse nature.

Qui et al. demonstrated the suitability of nitrogen-modified carbon material as support of PdAg catalysts for the decomposition of FA [67]. In that case, nitrogen modified reduced graphene oxide (N-rGO) was used and catalysts with diverse nanoparticle compositions (i.e. Pd$_2$Ag$_1$/N-rGO, Pd$_1$Ag$_1$/N-rGO, Pd$_1$Ag$_2$/N-rGO, Pd$_1$Ag$_4$/N-rGO, Pd$_1$Ag$_6$/N-rGO, Pd$_1$Ag$_8$/N-rGO) were prepared from AgNO$_3$ and Pd(NO$_3$)$_2$. The analysis of the characterisation results suggested that Pd atoms were more prone to be at the surface of the nanoparticles, which was not explained by the standard electrode potential of the pairs Pd^{2+}/Pd and Ag^+/Ag (0.951 V and 0.800 V, vs. SHE, respectively). Then, such Pd surface enrichment was related to the reducing ability of N-rGO for pre-forming Ag particles. Interestingly, it was shown in that study that low Pd-loading catalyst (Pd_1Ag_6/N-rGO) was able to achieve a relatively high TOF value (171 h^{-1}) at room temperature. Etemadi el at. reported on the utilisation of Ag-core Pd-shell nanoparticles supported on nitrogen-doped graphene carbon nanotube aerogel (Ag@Pd/N-GCNT) [69]. In that study graphene and CNTs were integrated into a hybrid aerogel, which was subsequently loaded with Ag and Pd in consecutive steps, to obtain controlled composition core-shell nanoparticles (i.e. Ag_1@Pd_2/N-GCNT, Ag_1@Pd_1/N-GCNT, Ag_2@Pd_1/N-GCNT, Ag/N-GCNT, and Pd/N-GCNT). It was observed that, aside from the N-sites, graphene carbon nanotube aerogel had additional stabilisation points for the nanoparticles that prevent their aggregation. In that case, catalysts with a Ag/Pd molar ratio 1/1 displayed the best activity among the investigated composition, with a TOF value of 413 h^{-1} at 25 °C.

Wang et al. analysed the catalytic performance of AgPd nanoparticles loaded on N-doped porous carbon synthesised from the carbonisation of a zeolitic imidazolate framework (ZIF-8) at various temperatures (i.e. 800, 900, and 1000 °C) [86]. The schematic illustration of the steps followed in the preparation of AgPd-based catalysts supported on nanoporous carbon (AgPd@NPC) is depicted in Figure 6.

The analysis of the composition of the catalysts indicated that the final Zn content decreased as the carbonisation temperature increased (7.49, 1.69, and 0.55 wt.% for nanoporous carbon obtained at 800, 900 and 1000 °C, respectively). XPS analysis demonstrated that there was an important interaction and electron transfer between Zn and alloy AgPd nanoparticles. The catalytic performance was assessed while checking the effect of both the composition of the nanoparticles and the temperature used in the carbonisation of carbon precursor. It was observed that both parameters played key roles. On the one hand, it was shown that, for a fixed carbonisation temperature, the catalyst with a composition of Ag_1Pd_4 displayed the most promising behaviour. On the other hand, the comparison of catalysts with Ag_1Pd_4 and loaded on carbon prepared at different temperatures revealed that the nanoporous carbon prepared at 900 °C was most suitable in this particular case. As a consequence of the optimisation of both parameters, Ag_1Pd_4@ZIF8-C (900) catalyst displayed the best performance among those investigated with a 100% of selectivity and a TOF value of 936 h^{-1} at 80 °C. Furthermore, that catalyst showed outstanding recyclability during five consecutive reaction runs.

Yamashita et al. also reported on the investigation of PdAg catalysts supported on N-containing carbon material [15]. Their study was centred in the development of efficient dual catalysts towards the reversible delivery/storage of H_2 in FA/CO_2. For that, PdAg nanoparticles loaded on phenylamine-functionalised mesoporous carbon (PdAg/amine-MSC) were synthesised from $Pd(NO_3)_2$ and $AgNO_3$. The experimental approach followed in that study produced amine-MSC with a concentration of amine groups of around 0.57 mmol g^{-1}. Reference samples were also prepared (Pd/MSC, Pd/amine-MSC, and PdAg/MSC). The characterisation of PdAg/amine-MSC revealed the presence of bimetallic alloy nanoparticles with an average size of 1.2 nm and homogeneous distribution of both elements. It was also suggested that metal nanoparticles were loaded near the amine groups and that such functionalisation favoured the formation of Ag-core rich and Pd-shell rich nanoparticles. As a result of such features, PdAg/amine-MSC showed an excellent catalytic activity with a TOF value as high as 5638 h^{-1} at 75 °C, which was much superior to that value obtained for the reference catalysts. The role of phenylamine in the catalytic performance was investigated by DFT calculations, which revealed that both reaction steps, O-H dissociation and H_2 desorption were boosted by such N-groups.

Wang et al. recently addressed a study in which the importance of the support morphology was investigated by preparing AgPd catalysts supported on N-decorated porous carbon nanosheets prepared from glucose and g-C_3N_4 by a template-induced strategy [79]. The lamellar structure obtained served as support of AgPd nanoparticles with compositions of Ag_1Pd_9/NPC, Ag_3Pd_7/NPC, Ag_5Pd_5/NPC, Ag_7Pd_3/NPC, Ag_9Pd_1/NPC, Pd/NPC, and Ag/NPC. Among them, Ag_1Pd_9/NPC was the most active catalyst. It was observed that the amount of template (g-C_3N_4) used in the synthesis did not have any effect on the catalysis of the resulting materials, while the morphology and surface properties of the support (which were varied by controlling the synthesis conditions) had a pivotal role in the final performance of the catalysts.

3.3. Metal Oxide-containing Carbon-based PdAg Catalysts

The presence of metal oxides can provide a basic character to carbon-supported catalysts. Also, it is known that certain metal oxides may enhance the final catalytic performance by serving as sites for the preferable adsorption of CO, and/or modifying the electronic properties of the metal nanoparticles.

Xu's group reported on the utilisation of multicomponent PdAg catalysts with rGO, in which the presence of a metal oxide benefited the formation of electron-rich PdAg species [32]. In that case, zirconia/porous carbon/reduced graphene oxide (ZrO_2/C/rGO) nanocomposites derived from MOF (UiO-66)/GO were synthesised by using the experimental approach illustrated in Figure 3. The effect of the composition of the nanoparticles was assessed by preparing catalysts with different Pd to Ag molar ratios (i.e. $Pd_{0.9}Ag_{0.1}$, $Pd_{0.7}Ag_{0.3}$, $Pd_{0.6}Ag_{0.4}$, $Pd_{0.5}Ag_{0.5}$, $Pd_{0.4}Ag_{0.4}$, and $Pd_{0.2}Ag_{0.8}$), as well as the monometallic counterparts. The evaluation of the catalytic activity at 60 °C suggested that, under the experimental conditions used in that study, the optimum composition of the nanoparticles was $Pd_{0.6}Ag_{0.4}$ and the further addition of Pd resulted in a lower activity. The best-performing catalysts among the investigated ($Pd_{0.6}Ag_{0.4}$@ZrO_2/C/rGO) showed a TOF value of 4500 h^{-1} (at 60 °C) and 100% selectivity towards the dehydrogenation reaction. Such outstanding performance was ascribed to the small size and high dispersion of the nanoparticles as well as the electron-rich PdAg surface species.

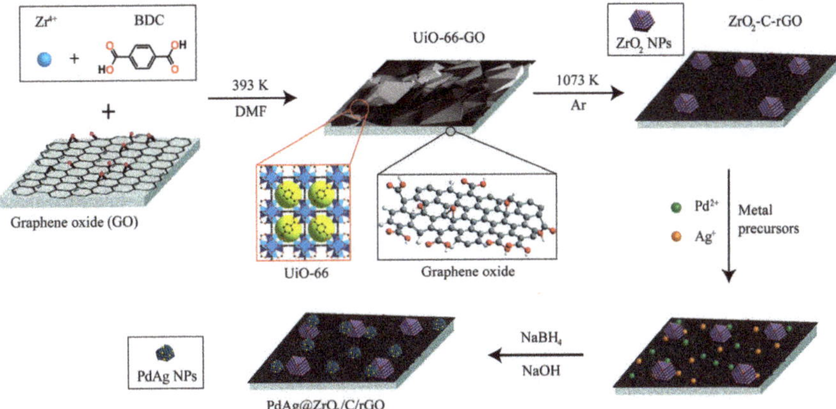

Figure 3. Schematic illustration for preparation of the PdAg@ZrO_2/C/rGO nanocatalyst. Reprinted with permission from [32].

Following Xu's work [32], Wang et al. studied the performance of PdAg catalysts supported on N-doped porous carbon embedded ZrO_2 nanohybrids obtained from UiO materials after calcination at temperatures ranging from 700 to 1000 °C (PdAg@ZrO_2/C) [81]. XRD analysis was used to determine the phase of zirconia present for each synthesis temperature (i.e. amorphous at 700 °C, and monoclinic (m-ZrO_2) and tetragonal (t-ZrO_2) at higher temperatures). Additional catalysts with various supports (t-ZrO_2, carbon-free ZrO_2, pure carbon black, N-doped porous carbon, and N-free ZrO_2/C) were prepared to get insight into the role of ZrO_2. Detailed characterisation of the catalysts revealed that there were oxygen vacancies in ZrO_2, which were formed due to the oxygen transfer from ZrO_2 to N-carbon. Furthermore, the relative basicity of the catalysts was also investigated, and the results obtained indicated that the catalysts prepared by carbonisation at 900 °C had the largest content of basic sites. Such observation was consistent with its better catalytic performance as compared to the rest of the samples under evaluation. The enhanced activity displayed by ZrO_2-containing catalysts was ascribed to the role of that oxide in both modifying the electronic properties of metal nanoparticles by creating surface electron-rich PdAg nanoparticles, as well as in providing basicity to the final materials.

Recently, Lu et al. designed carbon-supported metal oxide-containing PdAg catalysts with outstanding catalytic performance [72]. In that case, the advantage of the basic character of CeO_2 and interface properties optimisation by the strong coordination between that oxide and metals was taken into account to fabricate efficient $PdAg$-CeO_2 catalysts supported on mesoporous carbon ($PdAg$-CeO_2/MC). The mesoporosity of the selected carbon was claimed to benefit the mass transfer process and increase the reaction kinetics. Catalysts were obtained from a surfactant-free co-reduction approach using Na_2PdCl_4, $AgNO_3$, and $Ce(NO_3)_3 \cdot 6H_2O$, as metal precursors, and $NaBH_4$ as a reducing agent. Both Pd to Ag ratio (i.e. $Pd_{0.9}Ag_{0.1}$, $Pd_{0.7}Ag_{0.3}$, $Pd_{0.4}Ag_{0.4}$, and $Pd_{0.5}Ag_{0.5}$) and CeO_2 to carbon ratio were varied. In addition, carbon-free and CeO_2-free catalysts were also assessed as reference materials. According to the characterisation results, $PdAg$-CeO_2 composite was formed on the surface of the carbon material. XPS analysis confirmed the presence of Ce^{3+}, revealing the formation of oxygen vacancies that serve as defects to coordinate with the carbon support and metals. Furthermore, the comparison with CeO_2-free catalyst revealed that the presence of such oxide decreased the crystallinity of PdAg nanoparticles, thus resulting in the formation of more active site on the surface of the nanoparticles. The resulting catalysts were assessed towards the dehydrogenation of FA using sodium formate (SF, HCOONa) as an additive (see Figure 4).

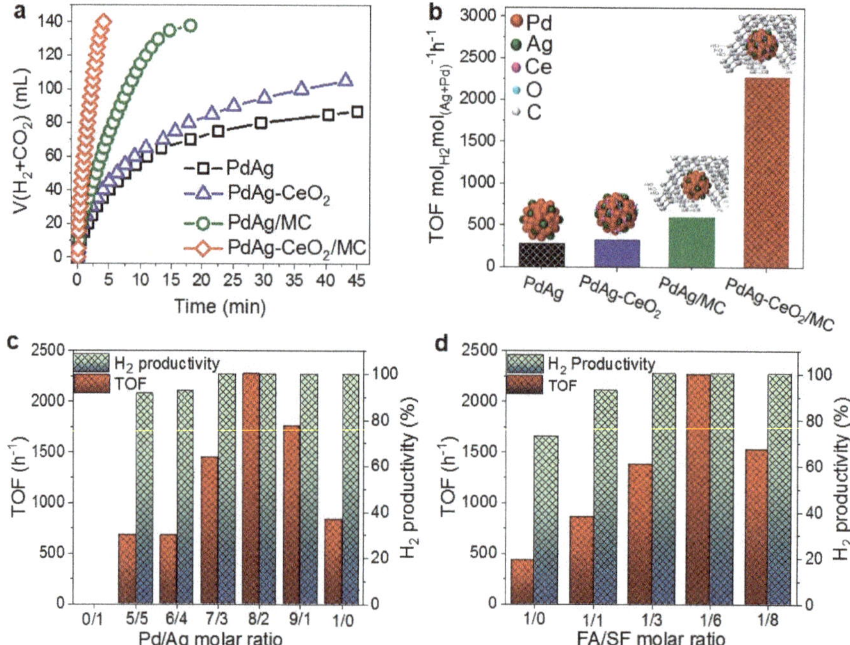

Figure 4. (**a**) Volume of the generated gas (H_2 + CO_2) versus time and (**b**) the corresponding TOF values for FA dehydrogenation in a FA-SF solution at 30 °C over PdAg, PdAg-CeO_2, PdAg/MC and PdAg-CeO_2/MC catalysts, respectively (n_{FA}/n_{SF} = 1/6; $n_{(Pd+Ag)}$:n_{FA} = 0.033); (**c**) Catalytic performance of the PdAg-CeO_2/MC catalysts with different Pd/Ag molar ratios for FA dehydrogenation at 30 °C in a FA-SFs solution (n_{FA}/n_{SF} = 1/6; $n_{(Pd+Ag)}$: n_{FA} = 0.033); (**d**) Catalytic performance of the optimised $Pd_{0.8}Ag_{0.2}$-CeO_2/MC catalyst for FA dehydrogenation at 30 °C in FA-SF solutions with different FA/SF molar ratios. Reprinted with permission from [72].

Figure 4a depicts the gas evolution profiles of the studied systems, which, together with TOF values displayed in Figure 4b, points out the superiority of the multicomponent catalyst. PdAg-CeO$_2$/MC catalyst had a TOF of 2272.8 h^{-1}, which was more than 8 times that of PdAg and much higher than the sum of the individual components (i.e. PdAg-CeO$_2$ and PdAg/MC), which demonstrated the synergistic effect between mesoporous carbon and PdAg via modification with CeO$_2$. Additionally, the effect of the Pd to Ag ratio in the nanoparticles was investigated and, as observed in Figure 4c, the optimum value was found to be 8/2.

The beneficial role of CeO$_2$ in the decomposition of FA over carbon-based catalysts had been previously observed by other authors while investigating the use of Pd–Au/C and Pd–Ag/C catalysts [92]. In that study, the composition of the nanoparticles was firstly optimised by assessing the activity of Pd/C, Pd-Ag/C, Pd-Au/C, and Pd-Cu/C catalysts at 92 °C. The results of the gas evolution profiles indicated the superior performance of Pd-Ag/C and Pd-Au/C, which was related to the weaker interaction of Ag and Au with CO molecules. In that case, even though an enhanced activity was observed upon addition of CeO$_2$, the reason for such improvement was not deeply analysed.

A more complex system was designed by Kim et al. [75]. In that study, bimetallic PdAg catalysts loaded on Fe$_3$O$_4$/nitrogen-doped reduced graphene oxide (PdAg/Fe$_3$O$_4$/N-rGO) were prepared. The motivation of the preparation of such multicomponent catalysts was to obtain a dual-functional catalyst for the generation of H$_2$ from FA and the subsequent utilisation of the generated H$_2$ for the selective defunctionalisation of lignin-derived chemicals. The incorporation of N heteroatoms was claimed to favour the anchoring of the nanoparticles and promote the dehydrogenation reaction, while Fe$_3$O$_4$ was incorporated to achieve easy separation of the catalysts. The final catalysts were deeply characterised by means of microscopic analysis, which confirmed the good dispersion of the metallic phases and the formation of PdAg alloys (See Figure 5). The resulting Pd$_{50}$Ag$_{50}$/Fe$_3$O$_4$/N-rGO displayed a TOF of 497 h^{-1} (at 50 °C) and CO was not detected.

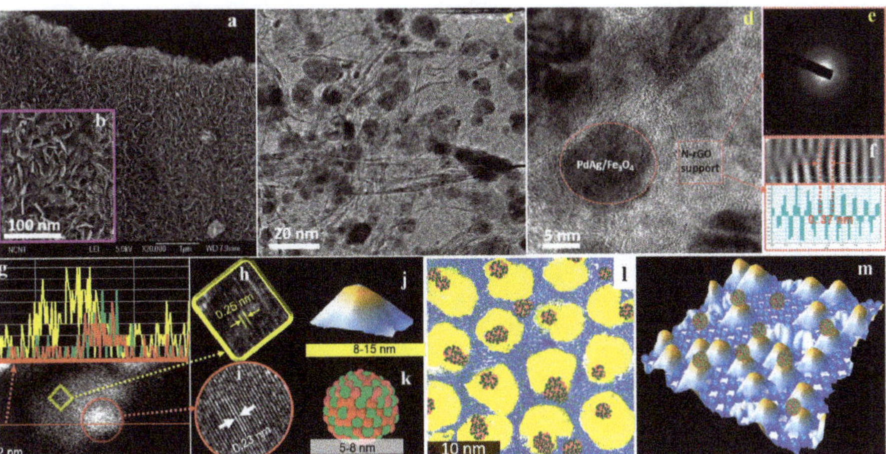

Figure 5. Microscopic analysis and proposed structure of Pd$_{50}$Ag$_{50}$/Fe$_3$O$_4$/N-rGO catalyst: (**a,b**) SEM images; (**c**) low-resolution TEM analysis; (**d**) high-resolution TEM analysis; (**e**) selected area electron diffraction pattern of N-rGO layer; (**f**) d spacing from stacked N-rGO layers; (**g**) dark field HAADF images with elemental line profile (yellow line, Fe; green line, Pd; red line, Ag); (**h,i**) CS-corrected TEM images of magnetite and PdAg nanoalloy; (**j,k**) proposed models for Fe$_3$O$_4$ and PdAg nanoalloy; (**l**) 2D scratch art model (yellow, magnetite; blue background, N-rGO; and green/red cluster, PdAg alloy); (**m**) proposed 3D structural model. Reprinted with permission from [75].

3.4. PdAg Catalysts Supported on Non-carbon Materials

Besides the catalyst supports based on carbon materials, some other interesting materials have been employed as support of PdAg nanoparticles to prepare catalysts for the dehydrogenation of FA.

Motivated by the excellent performance achieved by those N-containing carbon-based catalysts, Lei et al. investigated the suitability of graphitic carbon nitride (g-C_3N_4) as support for the preparation of efficient PdAg-based catalysts [74]. Bimetallic AgPd nanoparticles were supported on g-C_3N_4, previously prepared from melamine, by impregnation of metal precursors in different proportions and subsequent reduction in liquid phase with $NaBH_4$, so that catalysts with various compositions of the nanoparticles (Ag_4Pd_{96}/g-C_3N_4, Ag_9Pd_{91}/g-C_3N_4, $Ag_{16}Pd_{84}/g$-C_3N_4, $Ag_{21}Pd_{79}/g$-C_3N_4, Ag/g-C_3N_4, and Pd/g-C_3N_4) were obtained. The characterisation of the samples confirmed the presence of small and homogeneously distributed alloy nanoparticles. Among analysed samples, Ag_9Pd_{91}/g-C_3N_4 exhibited the most promising performance with a TOF value as high as 480 h^{-1} at 50 °C and 100% of selectivity towards the production of H_2.

Jia et al. also reported on the use of g-C_3N_4-based catalysts [82]. In that case, g-C_3N_4, synthesised from urea, was impregnated with Ag and Pd precursors to obtain catalysts with the following compositions: Ag_1Pd_2/CN, Ag_1Pd_4/CN, Ag_1Pd_1/CN, Ag_2Pd_1/CN, Ag/CN, and Pd/CN. The characterisation of the catalysts indicated that, in that case, electron-deficient Pd species were present. Such species were attributed to the formation of covalent Pd-$N_{pyridinic}$ bonds. The results of the catalytic tests revealed that Ag_1Pd_2/CN was the most active sample and had a TOF value of 621 h^{-1} at 75 °C in absence of additives.

Wan et al. also addressed the use of g-C_3N_4 as a useful approach to afford highly efficient catalysts for the present application [77]. In that case, the preparation of the catalysts was *via* immobilisation of g-C_3N_4 within the channel of SBA-15 and subsequent deposition of AgPd nanoparticles with various Ag to Pd ratios (Ag/Pd ratios of 1:9, 2:8, 3:7, 1:0, and 0:1) by a co-reduction method. Furthermore, g-C_3N_4 content was also optimised. The results of the catalytic tests revealed that sample $Ag_{10}Pd_{90}/0.2CND/SBA$-15 displayed superior activity with a TOF value of 893 h^{-1} at 50 °C.

Despite their poor stability in water, the utilisation of MOF-based catalysts was also investigated for the decomposition of FA by synthesising immobilised or embedded bimetallic metal nanoparticles. Cheng et al. reported for the first time the use of MOF as support of catalysts for this application [84]. In that study, MIL-101 was selected due to its high specific surface area, high thermal stability, and high chemical stability in water. MIL-101 was loaded with metal nanoparticles of various compositions ($Ag_{20}Pd_{80}@MIL$-101, $Ag_{35}Pd_{65}@MIL$-101, $Ag_{48}Pd_{52}@MIL$-101, $Ag_{63}Pd_{37}@MIL$-101, $Ag_{78}Pd_{22}@MIL$-101, $Ag@MIL$-101, and $Pd@MIL$-101) by using H_2PdCl_4 and $AgNO_3$ as metal precursors, after which the crystallinity of the support was preserved. The results extracted from TEM and N_2 adsorption measurements revealed that the metal nanoparticles were immobilised into the cavities of the support. As expected, the activity of the catalysts was dependent on the composition of the nanoparticle, being optimum for $Ag_{20}Pd_{80}@MIL$-101 catalyst, which had a TOF value of 848 h^{-1} at 80 °C. In addition, the synergy between AgPd nanoparticles and the support was confirmed by comparison with reference catalysts.

The same research group also reported on the synergistic catalytic properties of AgPd nanoparticles encapsulated in the cages of ZIF-8 (AgPd@ZIF-8) by checking the activity of catalysts with different Ag/Pd ratio ($Ag_{18}Pd_{82}@ZIF$-8, $Ag_{25}Pd_{75}@ZIF$-8, $Ag_{48}Pd_{52}@ZIF$-8, $Ag_{58}Pd_{42}@ZIF$-8, $Ag_{76}Pd_{24}@ZIF$-8, $Ag@ZIF$-8, and $Pd@ZIF$-8) that were synthesised by solution infiltration of ZIF-8 with the metal precursors and subsequent treatment with $NaBH_4$ [83]. To ascertain the role of the support, $Ag_{18}Pd_{82}$ supported on carbon, SiO_2, and Al_2O_3 were also prepared. In that case, sample $Ag_{18}Pd_{82}@ZIF$-8 had a considerable enhanced catalytic performance as compared to the monometallic Pd analogue, while higher content of Ag in the nanoparticles resulted in poorer performances. Moreover, the recyclability tests revealed that $Ag_{18}Pd_{82}@ZIF$-8 was still active after five catalytic runs and that the integrity of the framework of ZIF-8 was preserved. It was postulated in that study that the results displayed by $Ag_{18}Pd_{82}$@ZIF-8 were a consequence of the strong molecular-scale synergy of AgPd alloy with small size (1.6 ± 0.2 nm).

Zhu et. al fabricated a core-shell AgPd@MOF catalyst based on MIL-100(Fe) and synthesised by a one-pot strategy [63]. The approach followed afforded the preparation of materials with uniform shape and controllable size, in which a uniform MIL-100(Fe) shell is coating the AgPd core. The synthetic method consisted in the formation of MIL-100(Fe) on the surface of the PVP-modified AgPd nanoparticles. The sizes of core and shell could be easily controlled by adjusting the concentration of the precursors used in the synthesis. In the case of AgPd nanoparticles, their sizes ranged from 14 to 86 nm, while the shell thickness varied from 7 to 118 nm. It was observed that the thickness of MIL-100(Fe) played an important role in the final catalytic performance. Thus, catalysts with the thinnest shell displayed the highest activity because of the fast mass transfer of FA molecules to the embedded metallic nanoparticles. Furthermore, the importance of the arrangement of AgPd nanoparticles and MIL-100(Fe) in a core-shell structure was also demonstrated. It was claimed that MIL-100(Fe) shell adsorbed FA molecules close to the active metal nanoparticles, favouring, therefore, the efficient contact between FA molecules and active sites that ultimately improved the catalytic performance.

A step further was taken by Gao et al., who explored the catalytic response of AgPd alloy nanoparticles encapsulated within an amine-functionalised UiO-66 (AgPd@NH_2-UiO-66) [85]. In that case, the nitrogen functionalisation served as anchoring points for Ag and Pd precursor, which were immobilised within the pores *via* electrostatic interaction with N-groups (See Figure 7). Catalysts with a molar ratio Ag:Pd of 1:4, 1:2, 1:1, 2:1, 4:1, 1:0 and 0:1, and high metal content (ranging from 3.80 to 13.10 wt. %, and from 3.40 to 12.59 wt. %, for Ag and Pd, respectively) were synthesised by following the same experimental approach. Among investigated, Ag_1Pd_4@NH_2-UiO-66 was the best-performing catalyst for the dehydrogenation of FA, with an initial TOF of 893 h^{-1} at 80 °C and without any additive. It was assumed in that study that the promising catalytic performance shown by the studied system was due to the good dispersion of small AgPd nanoparticles and the basic properties of amino groups present in the catalyst.

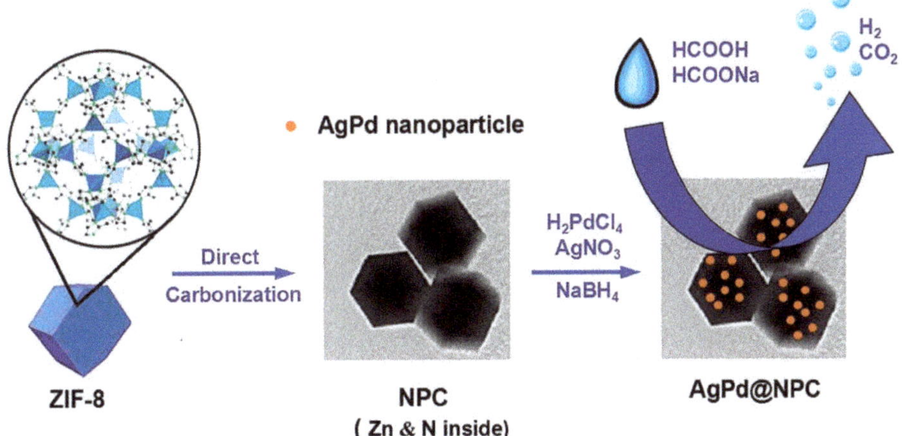

Figure 6. Schematic illustration for the preparation of AgPd@NPC. Reprinted with permission from [86].

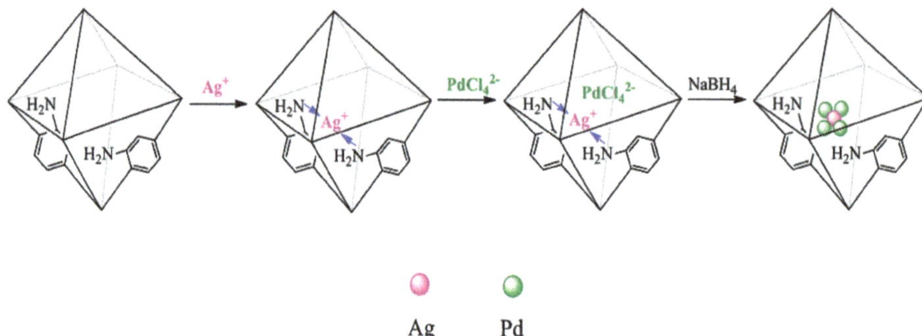

Figure 7. Schematic representation of the synthesis process of the AgPd@NH$_2$-UiO-66. Reprinted with permission from [85].

In an attempt to apply different AgPd-based catalytic structures for the reaction under study, Tsuji et al. investigated the catalytic activity of AgPd-core Pd-shell nanoparticles supported on TiO$_2$ prepared by a multistep method assisted by MW irradiation at different irradiation times (30 min, 1 h, or 2 h, for AgPd@Pd/TiO$_2$ (30 min), AgPd@Pd/TiO$_2$ (1 h), and AgPd@Pd/TiO$_2$ (2 h) catalysts, respectively) so as to have different extent of alloying of the AgPd core [93]. The characterisation of the samples revealed that all catalysts had very similar nanoparticle size, composition and morphology regardless of the heating time. XPS analysis confirmed the presence of Ag-Pd alloy in the interface of Ag-rich core and Pd-shell. Aside from that, it was also suggested that there was an electron transfer from the support to Ag and Pd. The dependence of the ability for the dehydrogenation of FA and the extent of alloying of the AgPd core was investigated with samples of various compositions Ag$_{100-x}$Pd$_x$ (x = 7, 10, and 15) @Pd/TiO$_2$ and a similar average nanoparticles size of 4.6 nm. It was observed that the best catalytic activity was displayed by Ag$_{93}$Pd$_7$@Pd/TiO$_2$.

Akbayrak recently reported on the utilisation of tungsten (VI) oxide, WO$_3$, which was selected due to its multiple oxidation states (that favour the electron transfer between support and nanoparticles) and strong metal-support interaction (SMSI) [76]. Catalysts with various compositions of the nanoparticles (Ag$_{0.25}$Pd/WO$_3$, Ag$_{0.5}$Pd/WO$_3$, Ag$_{0.75}$Pd/WO$_3$, Ag$_{1.0}$Pd/WO$_3$, Ag$_{1.5}$Pd/WO$_3$, Ag$_{0.25}$/WO$_3$) were prepared *via* a co-impregnation method. Among investigated, Ag$_{0.25}$Pd/WO$_3$ was the optimum catalyst but, despite the good activity shown in the first reaction run (TOF value of 683 h^{-1} at 50 °C), the second catalytic run evidenced its lack reusability due to aggregation of AgPd nanoparticles.

An et al. investigated the catalytic behaviour of amine-functionalized SBA-15 (SBA-15-Amine)-supported bimetallic AgPd nanoparticles [78]. According to the experimental methodology used in that study, 3-aminopropyltriethoxysilane (APTES) was used as amine source and the resulting SBA-15-Amine was loaded with Ag and Pd precursors in different proportions to obtain Pd/SBA-15-Amine, Ag$_1$Pd$_9$/SBA-15-Amine, Ag$_3$Pd$_7$/SBA-15-Amine, Ag$_5$Pd$_5$/SBA-15-Amine, Ag$_7$Pd$_3$/SBA-15-Amine, and Ag/SBA-15-Amine catalysts. It was observed that samples with moderate contents of Ag (i.e. Ag$_1$Pd$_9$/SBA-15-Amine, Ag$_3$Pd$_7$/SBA-15-Amine) showed enhanced performance as compared to Pd monometallic analogue, while higher Ag loading led to worsened activities. In particular, Ag$_1$Pd$_9$/SBA-15-Amine was shown to have the highest TOF among the investigated (964 h^{-1} at 50 °C), which was attributed to a molecular-scale synergistic effect of the AgPd alloy.

Zahmakiran et al. reported on a catalytic architecture based on bimetallic PdAg alloy and MnO$_x$ nanoparticles supported on amine-grafted silica (PdAg-MnOx/N-SiO$_2$) prepared from Pd(NO$_3$)·2H$_2$O, AgNO$_3$, and Mn(NO$_3$)$_2$·4H$_2$O as metal precursors, and 3-triethoxysilylpropylamine as the nitrogen source [62]. That investigation was based on a previous study of the same research group, in which the positive effect of MnO$_x$ as CO "trap" was observed [94]. Lately, they reported more complex catalysts based on bimetallic nanoparticles by using the same concept as that used for the monometallic system. Among the investigated compositions, catalysts with a composition of Pd$_{0.44}$Ag$_{0.19}$Mn$_{0.37}$ (i.e.,

1.27, 0.57, and 0.54 wt. % of Pd, Ag, and Mn, respectively) displayed the best catalytic performance. Furthermore, the comparison with the monometallic and binary counterpart catalysts confirmed the pivotal role of Ag in avoiding deactivation of the catalysts due to the generation of poison reaction by-products. In addition, control samples were also prepared by a physical mixture of the individual components (i.e. $Pd/N-SiO_2$, $Ag/N-SiO_2$, $Mn/N-SiO_2$), whose inferior catalytic activity confirmed that the synergistic effect observed for $Pd_{0.44}Ag_{0.19}-Mn_{0.37}/N-SiO_2$ is linked to the proximity of the components. A remarkable aspect of that catalytic system, is that it displayed total conversion of FA and high selectivity at room temperature and in absence of additives, and it retained 86% of the initial activity after five consecutive reaction runs. The effect of grafted amine groups was also investigated, observing that, under the experimental conditions used, the optimum amine loading was 0.98 mmol NH_2/g. It was claimed that the positive effect of $-NH_2$ groups was related to the FA adsorption/storage process and the formation of PdAg and MnO_x nanoparticles on the surface of the support. In addition, the beneficial effect of MnO_x was related to its ability to serve as a "pinning/anchoring site" for CO molecules, avoiding, therefore, the poisoning of PdAg catalytic active sites. Later, Wang et al. also incorporated MnO_x to the synthesis of the catalytic system used for the dehydrogenation of FA [80]. In that case, $AgPd-MnO_x$ supported on carbon nanospheres derived from biomass (i.e. sucrose) were investigated. It was observed that the catalysts displayed a promising activity and selectivity, with a TOF value of 3558 h^{-1} at 50 °C and with the addition of potassium formate (HCOOK) as an additive.

3.5. Photocatalytic Decomposition of Formic Acid

Recently, the decomposition of FA has also been tackled from the photocatalytic viewpoint. This approach is of great interest in the current energy scenario since it could join two aspects of the realisation of a sustainable energy future: the utilisation of sunlight, as a green and abundant energy source, and the production of hydrogen from a molecule that is potentially suitable for the carbon-neutral energy storage system. Despite its interest, few studies have been reported so far on the utilisation of photocatalytic systems based on PdAg bimetallic nanoparticles for the dehydrogenation of FA. Most of those studies address the use of the so-called Mott–Schottky heterojunctions, formed by a semiconductor support and metal nanoparticles, but some other strategies, such as the use of capping agents [95], can be also found in the literature.

Starting with TiO_2, as the photocatalytic material par excellence, Ago et al. investigated the catalysis of core-shell AgPd nanoparticles supported on TiO_2 ($AgPd@Pd/TiO_2$) [96]. As in the previous study reported by the same group, in which the production of H_2 was done by thermal decomposition of FA [93], the composition of metal nanoparticles was modified by dealloying the core of the nanoparticles for the preparation of $Ag_{100-x}Pd_x@Pd/TiO_2$ catalysts, with x = 7, 10, 15. In this case, the most promising composition of the nanoparticles was $Ag_{93}Pd_7@Pd/TiO_2$. Moreover, the effect of the support was also evaluated by using anatase (A) and the commercial TiO_2 P25 (P). The resulting materials had a Pd shell with a thickness of 0.8 nm for both A and P, and the morphology, size, and composition of the nanoparticles were similar regardless of the TiO_2 used. In order to monitor the photocatalytic response of the samples, gas evolution profiles were registered while irradiating the samples with a Xe lamp and heating at 27, 40, 60, 70, and 90 °C. For all the temperatures, the performance was superior for the illuminated samples, but some differences were observed for low and high temperatures. The analysis of the resulting data revealed that the initial reaction rate was enhanced by a factor of 1.5–1.6 upon irradiation of $AgPd@Pd/TiO_2$ (A) and $AgPd@Pd/TiO_2$ (P) at 27 °C, but such enhancement factor was 1.1–1.2 at 90 °C. The differences were claimed to depend on the migration of photogenerated electrons from TiO_2 to Pd. At low reaction temperatures, the electron transfer from the conduction band of TiO_2 to Pd shell takes place due to the difference in work function values (5.1, 4.7, and 4.0 eV for Pd, Ag, and TiO_2, respectively) and such electron-rich species are responsible for the enhancement of the catalytic performance (See Figure 8). Meanwhile, at higher temperatures, the electrons have higher migration rate, but the recombination of electron-hole pairs is favoured, leading to the reduction of the number of electrons on the Pd surface. Furthermore, the evaluation of the photocatalytic activity of AgPd@Pd/TiO$_2$ (A) and AgPd@Pd/TiO$_2$ (P) revealed the better performance of the anatase-based catalyst, which was attributed to the higher specific surface area and strong interaction of anatase phase with AgPd@Pd particles as well as to the slower electron-hole recombination compared to rutile phase of P25.

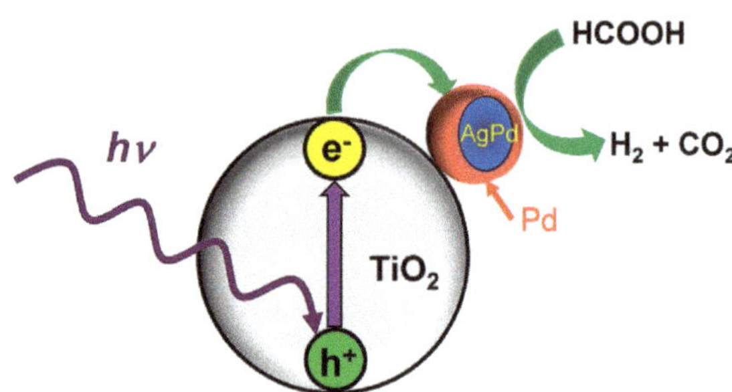

Figure 8. Enhancement mechanism of catalytic activity of AgPd@Pd/TiO$_2$ under UV and Vis photoirradiation. Reprinted with permission from [96].

Visible light-assisted photodecomposition of FA has also been investigated. Stucky et al. reported their study on AgPd bimetallic nanocrystals supported on *g*-C$_3$N$_4$ [97]. That support was not only selected considering its ability for anchoring metal nanoparticles and assisting the dehydrogenation reaction, but also because of the favourable electron transfer to the metal active phase. In that case, mesoporous hollow C$_3$N$_4$ spheres were used due to their better photoelectric features and higher surface area compared with conventional C$_3$N$_4$. Nanoparticles with a Ag:Pd ratio of 1:1 and an average size of 7.5 ± 1.0 nm were synthesised, and the photocatalytic ability of the resulting materials towards the H$_2$ production from FA was assessed. The comparison of that catalyst with reference samples, based on activated carbon and silica, pointed out the importance of the support in achieving well-dispersed AgPd nanocrystals. XPS analysis also confirmed the electron donation from surface unsaturated nitrogen atoms of the support to Pd. The formation of such surface electron-rich Pd species can be further promoted upon illumination *via* photoelectron transfer from the support, which ultimately suppresses the electron-hole pair recombination. As a consequence of such features, the developed catalyst showed excellent ability towards the dehydrogenation of FA under visible light irradiation. A set of photocatalysts with similar nanoparticle size and same metal loading was prepared to evaluate the effect of the composition of the nanoparticles (Ag3Pd, AgPd, AgPd3, and Pd). It was observed that the catalyst with a composition Ag:Pd of 1:1 displayed the highest TOF among analysed samples (91 and 254 h^{-1}, under dark and visible light irradiation at 30 °C, respectively), which was much higher than those values calculated for the monometallic analogue catalyst (48 and 137 h^{-1}, respectively).

Yu et. al. recently addressed the fabrication of Mott–Schottky heterojunctions constituted by PdAg nanowires (NWs) on *g*-C$_3$N$_4$ [98]. Bimetallic NWs with various Pd/Ag ratio were formed *in-situ* on the support to fabricate photocatalysts with various compositions (Pd$_7$Ag$_3$ NWs@*g*-C$_3$N$_4$, Pd$_5$Ag$_5$ NWs@*g*-C$_3$N$_4$, Pd$_3$Ag$_7$ NWs@*g*-C$_3$N$_4$, and Pd NWs@*g*-C$_3$N$_4$). The characterisation of the resulting samples revealed that Pd species were electron-enriched due to the electron transfer from both Ag and *g*-C$_3$N$_4$. It was reported that O-H bond dissociation was favoured by the presence of electron-rich Pd species, as well as by the support, which acts as proton scavenger. As expected, the catalytic activity was strongly dependent on the composition of the nanoparticles, and initial TOF values of 346, 420, 242, and 105 h^{-1}, were achieved for Pd$_7$Ag$_3$ NWs@*g*-C$_3$N$_4$, Pd$_5$Ag$_5$ NWs@*g*-C$_3$N$_4$, Pd$_3$Ag$_7$ NWs@*g*-C$_3$N$_4$ and Pd NWs@*g*-C$_3$N$_4$, respectively, while monitoring the reaction under visible light irradiation and at 25 °C. Besides, Pd$_5$Ag$_5$ NWs@*g*-C$_3$N$_4$ was further used to study the effect of the concentration of the catalysts, the concentration of FA, reaction temperature, and visible light intensity, on the performance towards the H_2 production. The stability of the photocatalysts was also investigated by carrying out four consecutive photocatalytic runs, after which the initial activity of Pd_5Ag_5 NWs@g-C_3N_4 was preserved.

3.6. Theoretical Investigations

The supremacy of PdAg-catalysts over other compositions of the active metal phase has widely been demonstrated from experimental investigations but also by theoretical studies. As for many other reactions [99–102], density functional theory (DFT) calculations have been used to study the decomposition of FA. Huang et al. reported a DFT study for the decomposition of FA over noble metals (Pt, Au, Pd, etc.) in both liquid and gas phase [103]. The energy profiles revealed that Pd-based surfaces favoured the decomposition of FA in the gas phase. However, that model could not be used to elucidate the decomposition of FA in aqueous phase. To do that, the solvation model of metal was used to get more realistic insights and the authors assumed that CO_2 was dissolved, and it did not have any influence in surface reactions. Regarding H_2 evolution, they considered that homolytic Tafel reaction played a pivotal role in the decomposition of FA, and it was concluded that the Pd surface displayed the lowest global energy barrier. Additionally, the authors pointed out that such a model could be used to study bimetallic alloys. A model cluster with two layers of Pd and two layers of Ag was built to explain the enhanced behaviour of PdAg bimetallic systems as compared to Pd monometallic catalysts.

An interesting study about transition metal catalysts in the decomposition of FA was reported by Studt et al. [104]. That study dealt with calculations of the reaction energetics in the decomposition of FA on transition-metal surfaces (Ag, Cu, Pd, etc.) using DFT. Additionally, that model allowed to get insights into the kinetics of the decomposition of FA over the transition-metal catalysts. The authors considered three pathways to produce CO_2 and H_2, while only one route of dehydration was taken into consideration (Figure 9).

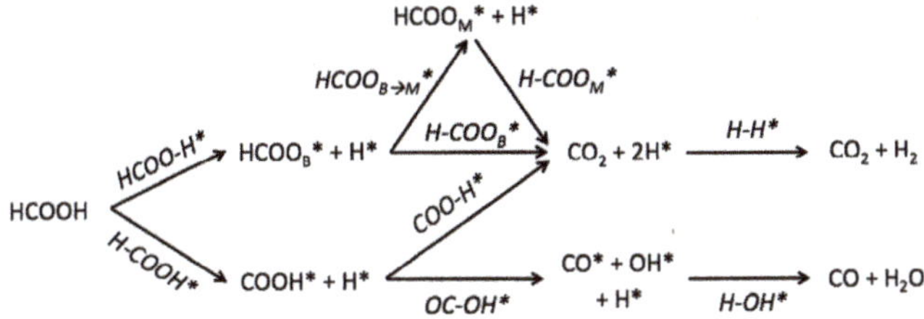

Figure 9. Reaction mechanism of FA decomposition. Reprinted with permission from [104].

All studied surfaces favoured the dehydrogenation rather than the dehydration pathway, and Pd surfaces displayed lower free-energy barriers than other transition metal surfaces. Other transition metal surfaces were explored by using the estimated adsorption energies of the species involved in the reaction over monometallic surfaces. From the TOFs values obtained in the decomposition of FA towards H_2 and CO_2, it was observed that monometallic noble metals did not display the best activity. It was found that Pd-based alloys with Au or Ag led to high selectivity to H_2 and CO_2 production. Moreover, the model could predict the low activity toward the CO formation, which is an extremely important factor for practical application in fuel cells. The selectivity towards either dehydrogenation or dehydration was studied by Ham et al [105]. The authors reported a theoretical study of PdAg-based catalysts for the decomposition of FA using the spin-polarized DFT to clarify which factors determine the selectivity of the reaction. The study was addressed for the evaluation of the impact of the Pd layer thickness over the enhancement of selectivity towards H_2 formation. They realised that the reaction rate of the dehydrogenation pathway depended on the number of Pd layers covering the Ag substrate, and the highest activity was attained for the lower thickness of the Pd layer. Furthermore, both lattice parameters and electronic structure in Pd-Ag catalysts were also discussed. It was observed that a charge transfer from Ag substrate to Pd monolayer favoured the dehydrogenation of FA, while strain effects partially hindered the H_2 production.

Cheng et al. recently published a comprehensive study towards the surface engineering structure of Pd-based alloy catalysts in which the relationship between the surface structure of the catalysts and their performance in the decomposition of FA was analysed by means of DFT and Sabatier analysis combined with experimental data [106]. In that case, not only PdAg alloys were analysed, but seven compositions of the nanoparticles (Pd–M; M = Ag, Au, Cu, Ni, Ir, Pt, and Rh) and five atomic arrangements (i.e. overlayer structure, core–shell structured Pd@M, core–shell structured Pd@Pd-M, uniform alloy, and a subsurface structure) were studied (see Figure 10). As a result of that analysis, two surface structures (i.e. $Pd_{2L}@Pd_1Ag_1$ and $Pd_{2L}@Pd_1Au_1$) were identified as potential good candidates to boost the selective dehydrogenation of FA.

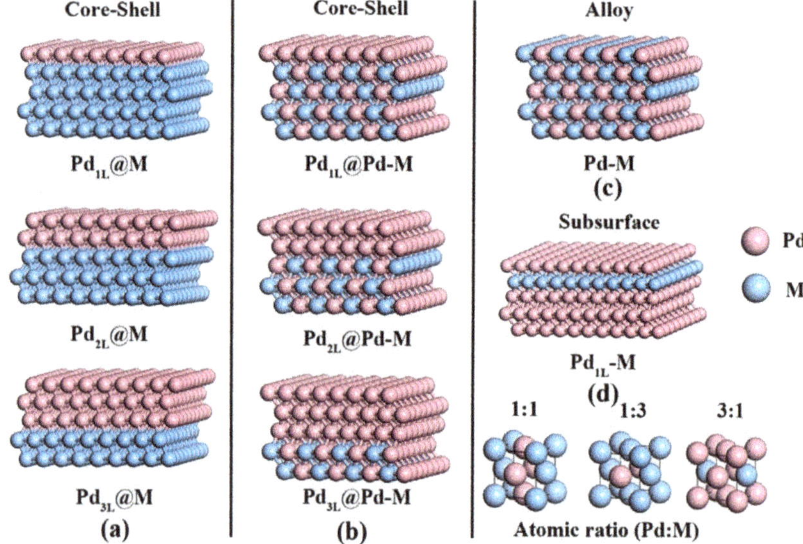

Figure 10. Models for the surface structures: (**a**) core–shell with Pd–M, (**b**) core–shell with a Pd alloy, (**c**) alloy, and (**d**) subsurface. Reprinted with permission from [106].

These studies evidence the importance of the theoretical outcomes for the optimisation of the composition and architecture of binary alloys based on Pd-Ag to obtain high-performance catalysts for the decomposition of FA.

4. Conclusions

The present review summarises some of the recent works on heterogeneous bimetallic PdAg catalytic systems for the production of H_2 from formic acid. The importance of the target reaction is contextualised by highlighting the crucial role of H_2 in the future energy scenario, as well as by pointing out the importance of LOHC to overcome the limitations displayed by conventional hydrogen technologies. The dehydrogenation of FA has widely been explored, while investigating diverse factors, such as the importance of the properties of the metal active phase and support, as well as the reaction conditions. In some other cases, the effect of the experimental conditions (i.e. concentration of FA and/or additives) has also been evaluated, but very little is known about the performance of the catalysts in high concentrated FA solutions. Theoretical investigations have also been conducted so as to ascertain the important characteristics to develop high-performance catalysts. Among those heterogeneous catalytic systems under investigation, the supremacy of Pd-based catalysts is undeniable. However, it is well-known at this point that the combination of Pd with a second metal provides enhanced catalytic performance due to the resulting electrical properties and resistance against CO poisoning effect. In particular, the use of PdAg-bimetallic systems offers outstanding results, which have motivated the search for multiple options for the optimisation of PdAg-bimetallic systems while getting insight into the role of the composition and structure of the nanoparticles, the properties of the catalytic support, the development of photocatalytic systems, and so forth. Concerning the properties of the metal active phase, the need for developing cost-effective catalysts has boosted the search for new metal nanoarchitectures with a larger fraction of surface Pd active sites. It has also been demonstrated that the incorporation of basic nitrogen functionalities into the support can greatly enhance the catalytic performance by two main effects: i) serving as anchoring points of metal active sites and avoiding the aggregation of the nanoparticles; ii) favouring the dissociation of the O-H bond of FA molecules. Moreover, the presence of N-containing groups has been related in some cases to the tunable reducibility of Ag and Pd, hence contributing to the surface engineering of the bimetallic structures. It has also been observed that the use PdAg-based photocatalysts are a promising alternative to afford suitable activity and selectivity values at room or moderate temperatures. However, few reports on that approach have been reported so far.

Despite the efforts devoted so far towards the design and development of PdAg-based bimetallic catalysts for the production of H_2 from FA, there are still some major shortcomings that limit their practical application and that should be the focus of future investigations. One of them is the need for extra additives, generally in the form of formate (HCOONa or HCOOK), as well as high temperatures, that are detrimental to the practical application of FA as a H_2 storage molecule. Developing catalytic systems able to produce H_2 from additive-free reaction medium while displaying high activity and selectivity is highly desirable. Another weak point of most catalysts developed so far is their stability, since an important decay is normally observed after a few consecutive catalytic runs. The deactivation issue is frequently related with changes in the properties of the nanoparticles, in terms of size and/or electronic properties, as well as with leaching of the metal active phase, but some other aspects such as the adsorption of reaction intermediates and CO poisoning have been also observed in some studies. It has been shown in some cases that the deactivation of the catalysts is less marked for bimetallic PdAg-based catalysts than for the monometallic counterpart, but this aspect is still focus of improvement. It could be envisaged that an interesting alternative could be the use of advanced encapsulated metal catalysts, in which the active phase is protected from leaching and/or surface modification while securing the mass transfer to and from its surface. Great achievements have recently been reported while investigating on the catalytic performance of PdAg-based catalysts; however, there is still a lot of work to be done to reach the maturity level accomplished with monometallic Pd catalysts. It is expected that more research groups will be soon interested in the fascinating field of research related to the production of H_2 from hydrogen carrier molecules, so that the worldwide scientific community can join their efforts towards the drawing of a hopeful energy prospect.

Author Contributions: M.N.-G. designed the structure of the review and wrote the manuscript, D.S.-T. contributed to the writing of the manuscript, and D.C.-A. reviewed the paper. All authors approved the manuscript for publication.

Funding: The authors would like to thank Generalitat Valenciana (PROMETEO/2018/076 project) for the financial support. Furthermore, M.N.-G. gratefully acknowledges Generalitat Valenciana and Plan GenT (CDEIGENT/2018/027) for the postdoctoral grant. D.S.-T. thanks MICINN for a "Juan de la Cierva" postdoctoral contract (IJCI-2016-27636).

Conflicts of Interest: The authors declare no conflict of interest.

References

1. United States Environmental Protection Agency. Available online: https://www.epa.gov/ghgemissions/understanding-global-warming-potentials (accessed on 13 September 2019).

2. Arrhenius, S. XXXI. On the influence of carbonic acid in the air upon the temperature of the ground. *Lond. Edinb. Dublin Philos. Mag. J. Sci.* **1896**, *41*, 237–276. [CrossRef]

3. Hook, M.; Tang, X. Depletion of fossil fuels and anthropogenic climate change-A review. *Energy Policy* **2013**, *52*, 797–809. [CrossRef]

4. Abas, N.; Kalair, A.; Khan, N. Review of fossil fuels and future energy technologies. *Futures* **2015**, *69*, 31–49. [CrossRef]

5. Alvarez, A.; Bansode, A.; Urakawa, A.; Bavykina, A.V.; Wezendonk, T.A.; Makkee, M.; Gascon, J.; Kapteijn, F. Challenges in the Greener Production of Formates/Formic Acid, Methanol, and DME by Heterogeneously Catalyzed CO2 Hydrogenation Processes. *Chem. Rev.* **2017**, *117*, 9804–9838. [CrossRef]

6. Artz, J.; Muller, T.E.; Thenert, K.; Kleinekorte, J.; Meys, R.; Sternberg, A.; Bardow, A.; Leitner, W. Sustainable Conversion of Carbon Dioxide: An Integrated Review of Catalysis and Life Cycle Assessment. *Chem. Rev.* **2018**, *118*, 434–504. [CrossRef] [PubMed]

7. Yang, Y.; Ajmal, S.; Zheng, X.; Zhang, L. Efficient nanomaterials for harvesting clean fuels from electrochemical and photoelectrochemical CO2 reduction. *Sustain. Energy Fuels* **2018**, *2*, 510–537. [CrossRef]

8. Abdalla, A.M.; Hossain, S.; Nisfindy, O.B.; Azad, A.T.; Dawood, M.; Azad, A.K. Hydrogen production, storage, transportation and key challenges with applications: A review. *Energy Convers. Manag.* **2018**, *165*, 602–627. [CrossRef]

9. Berenguer-Murcia, A.; Marco-Lozar, J.P.; Cazorla-Amoros, D. Hydrogen Storage in Porous Materials: Status, Milestones, and Challenges. *Chem. Rec.* **2018**, *18*, 900–912. [CrossRef]

10. Jang, M.; Jo, Y.S.; Lee, W.J.; Shin, B.S.; Sohn, H.; Jeong, H.; Jang, S.C.; Kwak, S.K.; Kang, J.W.; Yoon, C.W. A High-Capacity, Reversible Liquid Organic Hydrogen Carrier: H_2 —Release Properties and an Application to a Fuel Cell. *ACS Sustain. Chem. Eng.* **2019**, *7*, 1185–1194. [CrossRef]

11. Niermann, M.; Drunert, S.; Kaltschmitt, M.; Bonhoff, K. Liquid organic hydrogen carriers (LOHCs)—Techno-economic analysis of LOHCs in a defined process chain. *Energy Environ. Sci.* **2019**, *12*, 290–307. [CrossRef]

12. Modisha, P.M.; Ouma, C.N.M.; Garidzirai, R.; Wasserscheid, P.; Bessarabov, D. The Prospect of Hydrogen Storage Using Liquid Organic Hydrogen Carriers. *Energy Fuels* **2019**, *33*, 2778–2796. [CrossRef]

13. Preuster, P.; Papp, C.; Wasserscheid, P. Liquid Organic Hydrogen Carriers (LOHCs): Toward a Hydrogen-free Hydrogen Economy. *Acc. Chem. Res.* **2017**, *50*, 74–85. [CrossRef] [PubMed]

14. Enthaler, S.; Von Langermann, J.; Schmidt, T. Carbon dioxide and formic acid—The couple for environmental-friendly hydrogen storage? *Energy Environ. Sci.* **2010**, *3*, 1207–1217. [CrossRef]

15. Masuda, S.; Mori, K.; Futamura, Y.; Yamashita, H. PdAg Nanoparticles Supported on Functionalized Mesoporous Carbon: Promotional Effect of Surface Amine Groups in Reversible Hydrogen Delivery/Storage Mediated by Formic Acid/CO2. *ACS Catal.* **2018**, *8*, 2277–2285. [CrossRef]

16. Salinas-Torres, D.; Navlani-Garcia, M.; Mori, K.; Kuwahara, Y.; Yamashita, H. Nitrogen-doped carbon materials as a promising platform toward the efficient catalysis for hydrogen generation. *Appl. Catal. A Gen.* **2019**, *571*, 25–41. [CrossRef]

17. Eppinger, J.; Huang, K.-W. Formic Acid as a Hydrogen Energy Carrier. *ACS Energy Lett.* **2017**, *2*, 188–195. [CrossRef]

18. Coffey, R.S. The decomposition of formic acid catalysed by soluble metal complexes. *Chem. Commun.* **1967**, *18*, 923–924. [CrossRef]

19. Fellay, C.; Dyson, P.J.; Laurenczy, G. A viable hydrogen-storage system based on selective formic acid decomposition with a ruthenium catalyst. *Angew. Chem. Int. Ed.* **2008**, *47*, 3966–3968. [CrossRef]

20. Loges, B.; Boddien, A.; Junge, H.; Beller, M. Controlled generation of hydrogen from formic acid amine adducts at room temperature and application in H2/O2 fuel cells. *Angew. Chem. Int. Ed.* **2008**, *47*, 3962–3965. [CrossRef]

21. Iglesias, M.; Oro, L.A. Mechanistic Considerations on Homogeneously Catalyzed Formic Acid Dehydrogenation. *Eur. J. Inorg. Chem.* **2018**, *2018*, 2125–2138. [CrossRef]

22. Grasemann, M.; Laurenczy, G. Formic acid as a hydrogen source—Recent developments and future trends. *Energy Environ. Sci.* **2012**, *5*, 8171. [CrossRef]

23. Li, Z.; Xu, Q. Metal-Nanoparticle-Catalyzed Hydrogen Generation from Formic Acid. *Acc. Chem. Res.* **2017**, *50*, 1449–1458. [CrossRef] [PubMed]

24. Onishi, N.; Iguchi, M.; Yang, X.; Kanega, R.; Kawanami, H.; Xu, Q.; Himeda, Y. Development of Effective Catalysts for Hydrogen Storage Technology Using Formic Acid. *Adv. Energy Mater.* **2019**, *9*. [CrossRef]

25. Sordakis, K.; Tang, C.; Vogt, L.K.; Junge, H.; Dyson, P.J.; Beller, M.; Laurenczy, G. Homogeneous Catalysis for Sustainable Hydrogen Storage in Formic Acid and Alcohols. *Chem. Rev.* **2018**, *118*, 372–433. [CrossRef] [PubMed]

26. Tedsree, K.; Li, T.; Jones, S.; Chan, C.W.A.; Yu, K.M.K.; Bagot, P.A.J.; Marquis, E.A.; Smith, G.D.W.; Tsang, S.C.E. Hydrogen production from formic acid decomposition at room temperature using a Ag-Pd core-shell nanocatalyst. *Nat. Nanotechnol.* **2011**, *6*, 302–307. [CrossRef] [PubMed]

27. Onishi, N.; Laurenczy, G.; Beller, M.; Himeda, Y. Recent progress for reversible homogeneous catalytic hydrogen storage in formic acid and in methanol. *Coord. Chem. Rev.* **2018**, *373*, 317–332. [CrossRef]

28. Mellmann, D.; Sponholz, P.; Junge, H.; Beller, M. Formic acid as a hydrogen storage material-development of homogeneous catalysts for selective hydrogen release. *Chem. Soc. Rev.* **2016**, *45*, 3954–3988. [CrossRef] [PubMed]

29. Hong, W.; Kitta, M.; Tsumori, N.; Himeda, Y.; Autrey, T.; Xu, Q. Immobilization of highly active bimetallic PdAu nanoparticles onto nanocarbons for dehydrogenation of formic acid. *J. Mater. Chem. A* **2019**, *7*, 18835–18839. [CrossRef]

30. Wang, Q.; Chen, L.; Liu, Z.; Tsumori, N.; Kitta, M.; Xu, Q. Phosphate-Mediated Immobilization of High-Performance AuPd Nanoparticles for Dehydrogenation of Formic Acid at Room Temperature. *Adv. Funct. Mater.* **2019**, *29*, 1903341. [CrossRef]

31. Wang, Q.; Tsumori, N.; Kitta, M.; Xu, Q. Fast Dehydrogenation of Formic Acid over Palladium Nanoparticles Immobilized in Nitrogen-Doped Hierarchically Porous Carbon. *ACS Catal.* **2018**, *8*, 12041–12045. [CrossRef]

32. Song, F.Z.; Zhu, Q.L.; Yang, X.; Zhan, W.W.; Pachfule, P.; Tsumori, N.; Xu, Q. Metal–Organic Framework Templated Porous Carbon-Metal Oxide/Reduced Graphene Oxide as Superior Support of Bimetallic Nanoparticles for Efficient Hydrogen Generation from Formic Acid. *Adv. Energy Mater.* **2018**, *8*, 1–5. [CrossRef]

33. Zhu, D.-J.; Wen, Y.-H.; Xu, Q.; Zhu, Q.-L.; Wu, X.-T. Surface-Amine-Implanting Approach for Catalyst Functionalization: Prominently Enhancing Catalytic Hydrogen Generation from Formic Acid. *Eur. J. Inorg. Chem.* **2017**, *2017*, 4808–4813. [CrossRef]

34. Li, Z.; Yang, X.; Tsumori, N.; Liu, Z.; Himeda, Y.; Autrey, T.; Xu, Q. Tandem Nitrogen Functionalization of Porous Carbon: Toward Immobilizing Highly Active Palladium Nanoclusters for Dehydrogenation of Formic Acid. *ACS Catal.* **2017**, *7*, 2720–2724. [CrossRef]

35. Mori, K.; Dojo, M.; Yamashita, H. Pd and Pd-Ag nanoparticles within a macroreticular basic resin: An efficient catalyst for hydrogen production from formic acid decomposition. *ACS Catal.* **2013**, *3*, 1114–1119. [CrossRef]

36. Martis, M.; Mori, K.; Fujiwara, K.; Ahn, W.S.; Yamashita, H. Amine-functionalized MIL-125 with imbedded palladium nanoparticles as an efficient catalyst for dehydrogenation of formic acid at ambient temperature. *J. Phys. Chem.C* **2013**, *117*, 22805–22810. [CrossRef]

37. Mori, K.; Naka, K.; Masuda, S.; Miyawaki, K.; Yamashita, H. Palladium Copper Chromium Ternary Nanoparticles Constructed In situ within a Basic Resin: Enhanced Activity in the Dehydrogenation of Formic Acid. *ChemCatChem* **2017**, *9*, 3456–3462. [CrossRef]

38. Navlani-Garcia, M.; Salinas-Torres, D.; Mori, K.; Kuwahara, Y.; Yamashita, H. Enhanced formic acid dehydrogenation by the synergistic alloying effect of PdCo catalysts supported on graphitic carbon nitride. *Int. J. Hydrogen Energy* **2018**. [CrossRef]

39. Navlani-Garcia, M.; Martis, M.; Lozano-Castello, D.; Cazorla-Amoros, D.; Mori, K.; Yamashita, H. Investigation of Pd nanoparticles supported on zeolites for hydrogen production from formic acid dehydrogenation. *Catal. Sci. Technol.* **2015**, *5*, 364–371. [CrossRef]

40. Navlani-Garcia, M.; Mori, K.; Wen, M.; Kuwahara, Y.; Yamshita, H. Size Effect of Carbon-Supported Pd Nanoparticles in the Hydrogen Production from Formic Acid. *Bull. Chem. Soc. Jpn.* **2015**, *1370*, 78–80. [CrossRef]

41. Mori, K.; Tanaka, H.; Dojo, M.; Yoshizawa, K.; Yamashita, H. Synergic Catalysis of PdCu Alloy Nanoparticles within a Macroreticular Basic Resin for Hydrogen Production from Formic Acid. *Chem. A Eur. J.* **2015**, *21*, 12085–12092. [CrossRef]

42. Navlani-Garcia, M.; Mori, K.; Nozaki, A.; Kuwahara, Y.; Yamashita, H. Investigation of Size Sensitivity in the Hydrogen Production from Formic Acid over Carbon-Supported Pd Nanoparticles. *ChemistrySelect* **2016**, *1*, 1879–1886. [CrossRef]

43. Navlani-Garcia, M.; Mori, K.; Nozaki, A.; Kuwahara, Y.; Yamashita, H. Screening of Carbon-Supported PdAg Nanoparticles in the Hydrogen Production from Formic Acid. *Ind. Eng. Chem. Res.* **2016**, *55*, 7612–7620. [CrossRef]

44. Garcia-Aguilar, J.; Navlani-Garcia, M.; Berenguer-Murcia, A.; Mori, K.; Kuwahara, Y.; Yamashita, H.; Cazorla-Amoros, D. Evolution of the PVP-Pd surface interaction in nanoparticles through the case study of formic acid decomposition. *Langmuir* **2016**, *32*, 12110–12118. [CrossRef] [PubMed]

45. Wen, M.; Mori, K.; Kuwahara, Y.; Yamashita, H. Plasmonic Au@Pd nanoparticles supported on a basic metal-organic framework: Synergic boosting of H2production from formic acid. *ACS Energy Lett.* **2017**, *2*, 1–7. [CrossRef]

46. Wu, Y.; Wen, M.; Navlani-Garcia, M.; Kuwahara, Y.; Mori, K.; Yamashita, H. Palladium nanoparticles supported on titanium doped graphitic carbon nitride for formic acid dehydrogenation. *Chem. Asian J.* **2017**, *12*, 860–867. [CrossRef]

47. Bulushev, D.A.; Jia, L.; Beloshapkin, S.; Ross, J.R.H. Improved hydrogen production from formic acid on a Pd/C catalyst doped by potassium. *Chem. Commun.* **2012**, *48*, 4184–4186. [CrossRef]

48. Bulushev, D.A.; Beloshapkin, S.; Plyusnin, P.E.; Shubin, Y.V.; Bukhtiyarov, V.I.; Korenev, S.V.; Ross, J.R.H. Vapour phase formic acid decomposition over PdAu/γ-Al$_2$O$_3$ catalysts: Effect of composition of metallic particles. *J. Catal.* **2013**, *299*, 171–180. [CrossRef]

49. Bulushev, D.A.; Sobolev, V.I.; Pirutko, L.V.; Starostina, A.V.; Asanov, I.P.; Modin, E.; Chuvilin, A.L.; Gupta, N.; Okotrub, A.V.; Bulusheva, L.G. Hydrogen production from formic acid over au catalysts supported on carbon: Comparison with Au catalysts supported on SiO$_2$ and Al$_2$O$_3$. *Catalysts* **2019**, *9*, 376. [CrossRef]

50. Jia, L.; Bulushev, D.A.; Podyacheva, O.Y.; Boronin, A.I.; Kibis, L.S.; Gerasimov, E.Y.; Beloshapkin, S.; Seryak, I.A.; Ismagilov, Z.R.; Ross, J.R.H. Pt nanoclusters stabilized by N-doped carbon nanofibers for hydrogen production from formic acid. *J. Catal.* **2013**, *307*, 94–102. [CrossRef]

51. Zacharska, M.; Podyacheva, O.Y.; Kibis, L.S.; Boronin, A.I.; Senkovskiy, B.V.; Gerasimov, E.Y.; Taran, O.P.; Ayusheev, A.B.; Parmon, V.N.; Leahy, J.J.; et al. Ruthenium Clusters on Carbon Nanofibers for Formic Acid Decomposition: Effect of Doping the Support with Nitrogen. *ChemCatChem* **2015**, *7*, 2910–2917. [CrossRef]

52. Zacharska, M.; Chuvilin, A.L.; Kriventsov, V.V.; Beloshapkin, S.; Estrada, M.; Simakov, A.; Bulushev, D.A. Support effect for nanosized Au catalysts in hydrogen production from formic acid decomposition. *Catal. Sci. Technol.* **2016**, *6*, 6853–6860. [CrossRef]

53. Bulushev, D.A.; Zacharska, M.; Shlyakhova, E.V.; Chuvilin, A.L.; Guo, Y.; Beloshapkin, S.; Okotrub, A.V.; Bulusheva, L.G. Single Isolated Pd2+ Cations Supported on N-Doped Carbon as Active Sites for Hydrogen Production from Formic Acid Decomposition. *ACS Catal.* **2016**, *6*, 681–691. [CrossRef]

54. Bulushev, D.A.; Zacharska, M.; Lisitsyn, A.S.; Podyacheva, O.Y.; Hage, F.S.; Ramasse, Q.M.; Bangert, U.; Bulusheva, L.G. Single Atoms of Pt-Group Metals Stabilized by N-Doped Carbon Nanofibers for Efficient Hydrogen Production from Formic Acid. *ACS Catal.* **2016**, *6*, 3442–3451. [CrossRef]

55. Zacharska, M.; Bulusheva, L.G.; Lisitsyn, A.S.; Beloshapkin, S.; Guo, Y.; Chuvilin, A.L.; Shlyakhova, E.V.; Podyacheva, O.Y.; Leahy, J.J.; Okotrub, A.V.; et al. Factors Influencing the Performance of Pd/C Catalysts in the Green Production of Hydrogen from Formic Acid. *ChemSusChem* **2017**, *10*, 720–730. [CrossRef] [PubMed]

56. Bulushev, D.A.; Zacharska, M.; Beloshapkin, S.; Guo, Y.; Yuranov, I. Catalytic properties of PdZn/ZnO in formic acid decomposition for hydrogen production. *Appl. Catal. A Gen.* **2018**, *561*, 96–103. [CrossRef]

57. Podyacheva, O.Y.; Bulushev, D.A.; Suboch, A.N.; Svintsitskiy, D.A.; Lisitsyn, A.S.; Modin, E.; Chuvilin, A.; Gerasimov, E.Y.; Sobolev, V.I.; Parmon, V.N. Highly Stable Single-Atom Catalyst with Ionic Pd Active Sites Supported on N-Doped Carbon Nanotubes for Formic Acid Decomposition. *ChemSusChem* **2018**, *11*, 3724–3727. [CrossRef]

58. Navlani-Garcia, M.; Mori, K.; Kuwahara, Y.; Yamashita, H. Recent strategies targeting efficient hydrogen production from chemical hydrogen storage materials over carbon-supported catalysts. *NPG Asia Mater.* **2018**, *10*, 277–292. [CrossRef]

59. Navlani-Garcia, M.; Mori, K.; Salinas-Torres, D.; Kuwahara, Y.; Yamashita, H. New Approaches Toward the Hydrogen Production From Formic Acid Dehydrogenation Over Pd-Based Heterogeneous Catalysts. *Front. Mater.* **2019**, *6*, 44. [CrossRef]

60. Lee, S.; Cho, J.; Jang, J.H.; Han, J.; Yoon, S.P.; Nam, S.W.; Lim, T.H.; Ham, H.C. Impact of d-Band Occupancy and Lattice Contraction on Selective Hydrogen Production from Formic Acid in the Bimetallic Pd3M (M = Early Transition 3d Metals) Catalysts. *ACS Catal.* **2016**, *6*, 134–142. [CrossRef]

61. Choi, B.-S.; Song, J.; Song, M.; Goo, B.S.; Lee, Y.W.; Kim, Y.; Yang, H.; Han, S.W. Core–Shell Engineering of Pd–Ag Bimetallic Catalysts for Efficient Hydrogen Production from Formic Acid Decomposition. *ACS Catal.* **2019**, *9*, 819–826. [CrossRef]

62. Bulut, A.; Yurderi, M.; Karatas, Y.; Say, Z.; Kivrak, H.; Kaya, M.; Gulcan, M.; Ozensoy, E.; Zahmakiran, M. MnOx-Promoted PdAg Alloy Nanoparticles for the Additive-Free Dehydrogenation of Formic Acid at Room Temperature. *ACS Catal.* **2015**, *5*, 6099–6110. [CrossRef]

63. Ke, F.; Wang, L.; Zhu, J. An efficient room temperature core-shell AgPd@MOF catalyst for hydrogen production from formic acid. *Nanoscale* **2015**, *7*, 8321–8325. [CrossRef] [PubMed]

64. Huang, Y.; Xu, J.; Long, T.; Shuai, Q.; Li, Q. Reducing agent-structure-activity relationship of PdAg/C catalysts in formic acid decomposition for hydrogen generation. *J. Nanosci. Nanotechnol.* **2017**, *17*, 3798–3802. [CrossRef]

65. Ping, Y.; Yan, J.M.; Wang, Z.L.; Wang, H.L.; Jiang, Q. Ag0.1-Pd0.9/rGO: An efficient catalyst for hydrogen generation from formic acid/sodium formate. *J. Mater. Chem. A* **2013**, *1*, 12188–12191. [CrossRef]

66. Hu, C.; Mu, X.; Fan, J.; Ma, H.; Zhao, X.; Chen, G.; Zhou, Z.; Zheng, N. Interfacial effects in PdAg bimetallic nanosheets for selective dehydrogenation of formic acid. *ChemNanoMat* **2016**, *2*, 28–32. [CrossRef]

67. Huang, Y.; Xu, J.; Ma, X.; Huang, Y.; Li, Q.; Qiu, H. An effective low Pd-loading catalyst for hydrogen generation from formic acid. *Int. J. Hydrogen Energy* **2017**, *42*, 18375–18382. [CrossRef]

68. Jiang, Y.; Fan, X.; Xiao, X.; Qin, T.; Zhang, L.; Jiang, F.; Li, M.; Li, S.; Ge, H.; Chen, L. Novel AgPd hollow spheres anchored on graphene as an efficient catalyst for dehydrogenation of formic acid at room temperature. *J. Mater. Chem. A* **2015**, *4*, 657–666. [CrossRef]

69. Nabid, M.R.; Bide, Y.; Etemadi, B. Ag@Pd nanoparticles immobilized on a nitrogen-doped graphene carbon nanotube aerogel as a superb catalyst for the dehydrogenation of formic acid. *New J. Chem.* **2017**, *41*, 10773–10779. [CrossRef]

70. Yang, L.; Hua, X.; Su, J.; Luo, W.; Chen, S.; Cheng, G. Highly efficient hydrogen generation from formic acid-sodium formate over monodisperse AgPd nanoparticles at room temperature. *Appl. Catal. B Environ.* **2015**, *168–169*, 423–428. [CrossRef]

71. Feng, C.; Wang, Y.; Gao, S.; Shang, N.; Wang, C. Hydrogen generation at ambient conditions: AgPd bimetal supported on metal-organic framework derived porous carbon as an efficient synergistic catalyst. *Catal. Commun.* **2016**, *78*, 17–21. [CrossRef]

72. Zhang, Z.; Luo, Y.; Liu, S.; Yao, Q.; Qing, S.; Lu, Z.-H. A PdAg-CeO$_2$ nanocomposite anchored on mesoporous carbon: A highly efficient catalyst for hydrogen production from formic acid at room temperature. *J. Mater. Chem. A* **2019**, *7*, 21438–21446. [CrossRef]

73. Zhang, S.; Metin, O.; Su, D.; Sun, S. Monodisperse AgPd alloy nanoparticles and their superior catalysis for the dehydrogenation of formic acid. *Angew. Chem. Int. Ed.* **2013**, *52*, 3681–3684. [CrossRef] [PubMed]

74. Yao, F.; Li, X.; Wan, C.; Xu, L.; An, Y.; Ye, M.; Lei, Z. Highly efficient hydrogen release from formic acid using a graphitic carbon nitride-supported AgPd nanoparticle catalyst. *Appl. Surf. Sci.* **2017**, *426*, 605–611. [CrossRef]

75. Singh, A.K.; Jang, S.; Kim, J.Y.; Sharma, S.; Basavaraju, K.C.; Kim, M.-G.; Kim, K.-R.; Lee, J.S.; Lee, H.H.; Kim, D.-P. One-Pot Defunctionalization of Lignin-Derived Compounds by Dual-Functional Pd$_{50}$Ag$_{50}$/Fe$_3$O$_4$/N-rGO Catalyst. *ACS Catal.* **2015**, *5*, 6964–6972. [CrossRef]

76. Akbayrak, S. Decomposition of formic acid using tungsten(VI) oxide supported AgPd nanoparticles. *J. Colloid Interface Sci.* **2019**, *538*, 682–688. [CrossRef] [PubMed]

77. Xu, L.; Jin, B.; Zhang, J.; Cheng, D.-G.; Chen, F.; An, Y.; Cui, P.; Wan, C. Efficient hydrogen generation from formic acid using AgPd nanoparticles immobilized on carbon nitride-functionalized SBA-15. *RSC Adv.* **2016**, *6*, 46908–46914. [CrossRef]

78. Wan, C.; Yao, F.; Li, X.; Hu, K.; Ye, M.; Xu, L.; An, Y. Bimetallic AgPd Nanoparticles Immobilized on Amine-Functionalized SBA-15 as Efficient Catalysts for Hydrogen Generation from Formic Acid. *ChemistrySelect* **2016**, *1*, 6907–6913. [CrossRef]

79. Zhang, X.; Shang, N.; Shang, H.; Du, T.; Zhou, X.; Feng, C.; Gao, S.; Wang, C.; Wang, Z. Nitrogen-decorated porous carbon supported AgPd nanoparticles for boosting hydrogen generation from formic acid. *Energy Technol.* **2019**, *7*, 140–145. [CrossRef]

80. Zhang, X.; Shang, N.; Zhou, X.; Feng, C.; Gao, S.; Wu, Q.; Wang, Z.; Wang, C. AgPd-MnOxsupported on carbon nanospheres: An efficient catalyst for dehydrogenation of formic acid. *New J. Chem.* **2017**, *41*, 3443–3449. [CrossRef]

81. Feng, C.; Gao, S.; Shang, N.; Zhou, X.; Wang, C. Super Nanoteragonal ZrO$_2$ Embedded in Carbon as Efficient Support of PdAg Nanoparticle for Boosting Hydrogen Generation from Formic Acid. *Energy Technol.* **2018**, *6*, 2120–2125. [CrossRef]

82. Deng, Q.-F.; Xin, J.-J.; Ma, S.-K.; Cui, F.-J.; Zhao, Z.-L.; Jia, L.-H. Hydrogen Production from the Decomposition of Formic Acid over Carbon Nitride-Supported AgPd Alloy Nanoparticles. *Energy Technol.* **2018**, *6*, 2374–2379. [CrossRef]

83. Dai, H.; Xia, B.; Wen, L.; Du, C.; Su, J.; Luo, W.; Cheng, G. Synergistic catalysis of AgPd@ZIF-8 on dehydrogenation of formic acid. *Appl. Catal. B Environ.* **2015**, *165*, 57–62. [CrossRef]

84. Dai, H.; Cao, N.; Yang, L.; Su, J.; Luo, W.; Cheng, G. AgPd nanoparticles supported on MIL-101 as high performance catalysts for catalytic dehydrogenation of formic acid. *J. Mater. Chem. A* **2014**, *2*, 11060–11064. [CrossRef]

85. Gao, S.-T.; Liu, W.; Feng, C.; Shang, N.-Z.; Wang, C. A Ag-Pd alloy supported on an amine-functionalized UiO-66 as an efficient synergetic catalyst for the dehydrogenation of formic acid at room temperature. *Catal. Sci. Technol.* **2016**, *6*, 869–874. [CrossRef]

86. Feng, C.; Hao, Y.; Zhang, L.; Shang, N.; Gao, S.; Wang, Z.; Wang, C. AgPd nanoparticles supported on zeolitic imidazolate framework derived N-doped porous carbon as an efficient catalyst for formic acid dehydrogenation. *RSC Adv.* **2015**, *5*, 39878–39883. [CrossRef]

87. Jiang, K.; Xu, K.; Zou, S.; Cai, W.-B. B-doped pd catalyst: Boosting room-temperature hydrogen production from formic acid-formate solutions. *J. Am. Chem. Soc.* **2014**, *136*, 4861–4864. [CrossRef]

88. Navlani-Garcia, M.; Salinas-Torres, D.; Mori, K.; Kuwahara, Y.; Yamashita, H. Tailoring the Size and Shape of Colloidal Noble Metal Nanocrystals as a Valuable Tool in Catalysis. *Catal. Surv. Asia* **2019**, *23*, 127–148. [CrossRef]

89. Navlani-Garcia, M.; Salinas-Torres, D.; Mori, K.; Leonard, A.F.; Kuwahara, Y.; Job, N.; Yamashita, H. Insights on palladium decorated nitrogen-doped carbon xerogels for the hydrogen production from formic acid. *Catal. Today* **2019**, *324*, 90–96. [CrossRef]

90. Yu, Y.; Wang, X.; Liu, C.; Vladimir, F.; Ge, J.; Xing, W. Surface interaction between Pd and nitrogen derived from hyperbranched polyamide towards highly effective formic acid dehydrogenation. *J. Energy Chem.* **2020**, *40*, 212–216. [CrossRef]

91. Sun, J.; Qiu, H.; Cao, W.; Fu, H.; Wan, H.; Xu, Z.; Zheng, S. Ultrafine Pd Particles Embedded in Nitrogen-Enriched Mesoporous Carbon for Efficient H$_2$ Production from Formic Acid Decomposition. *ACS Sustain. Chem. Eng.* **2019**, *7*, 1963–1972. [CrossRef]

92. Zhou, X.; Huang, Y.; Xing, W.; Liu, C.; Liao, J.; Lu, T. High-quality hydrogen from the catalyzed decomposition of formic acid by Pd-Au/C and Pd-Ag/C. *Chem. Commun.* **2008**, 3540–3542. [CrossRef]

93. Hattori, M.; Shimamoto, D.; Ago, H.; Tsuji, M. AgPd@Pd/TiO$_2$ nanocatalyst synthesis by microwave heating in aqueous solution for efficient hydrogen production from formic acid. *J. Mater. Chem. A* **2015**, *3*, 10666–10670. [CrossRef]

94. Şener, T.; Demirci, U.B.; Gul, F.; Ata, A. Pd-MnO$_2$-Fe/C as electrocatalyst for the formic acid electrooxidation. *Int. J. Hydrogen Energy* **2015**, *40*, 6920–6926. [CrossRef]

95. Liu, H.; Huang, B.; Zhou, J.; Wang, K.; Yu, Y.; Yang, W.; Guo, S. Enhanced electron transfer and light absorption on imino polymer capped PdAg nanowire networks for efficient room-temperature dehydrogenation of formic acid. *J. Mater. Chem. A* **2018**, *6*, 1979–1984. [CrossRef]

96. Tsuji, M.; Shimamoto, D.; Uto, K.; Hattori, M.; Ago, H. Enhancement of catalytic activity of AgPd@Pd/TiO$_2$ nanoparticles under UV and visible photoirradiation. *J. Mater. Chem. A* **2016**, *4*, 14649–14656. [CrossRef]

97. Xiao, L.; Jun, Y.-S.; Wu, B.; Liu, D.; Chuong, T.T.; Fan, J.; Stucky, G.D. Carbon nitride supported AgPd alloy nanocatalysts for dehydrogenation of formic acid under visible light. *J. Mater. Chem. A* **2017**, *5*, 6382–6387. [CrossRef]

98. Liu, H.; Liu, X.; Yang, W.; Shen, M.; Geng, S.; Yu, C.; Shen, B.; Yu, Y. Photocatalytic dehydrogenation of formic acid promoted by a superior PdAg@g-C$_3$N$_4$ Mott–Schottky heterojunction. *J. Mater. Chem. A* **2019**, *7*, 2022–2026. [CrossRef]

99. Shan, N.; Zhou, M.; Hanchett, M.K.; Chen, J.; Liu, B. Practical principles of density functional theory for catalytic reaction simulations on metal surfaces—From theory to applications. *Mol. Simul.* **2017**, *43*, 861–885. [CrossRef]

100. Sameera, W.M.C.; Maseras, F. Transition metal catalysis by density functional theory and density functional theory/molecular mechanics. *Wiley Interdiscip. Rev. Comput. Mol. Sci.* **2012**, *2*, 375–385. [CrossRef]

101. Norskov, J.K.; Abild-Pedersen, F.; Studt, F.; Bligaard, T. Density functional theory in surface chemistry and catalysis. *Proc. Natl. Acad. Sci. USA* **2011**, *108*, 937–943. [CrossRef]

102. Singh, S.; Li, S.; Carrasquillo-Flores, R.; Alba-Rubio, A.C.; Dumesic, J.A.; Mavrikakis, M. Formic acid decomposition on Au catalysts: DFT, microkinetic modeling, and reaction kinetics experiments. *AIChE J.* **2014**, *60*, 1303–1319. [CrossRef]

103. Hu, C.; Ting, S.-W.; Chan, K.-Y.; Huang, W. Reaction pathways derived from DFT for understanding catalytic decomposition of formic acid into hydrogen on noble metals. *Int. J. Hydrogen Energy* **2012**, *37*, 15956–15965. [CrossRef]

104. Yoo, J.S.; Abild-Pedersen, F.; Norskov, J.K.; Studt, F. Theoretical analysis of transition-metal catalysts for formic acid decomposition. *ACS Catal.* **2014**, *4*, 1226–1233. [CrossRef]

105. Cho, J.; Lee, S.; Han, J.; Yoon, S.P.; Nam, S.W.; Choi, S.H.; Lee, K.-Y.; Ham, H.C. Importance of Ligand Effect in Selective Hydrogen Formation via Formic Acid Decomposition on the Bimetallic Pd/Ag Catalyst from First-Principles. *J. Phys. Chem. C* **2014**, *118*, 22553–22560. [CrossRef]

106. Yang, Y.; Xu, H.; Cao, D.; Zeng, X.C.; Cheng, D. Hydrogen Production via Efficient Formic Acid Decomposition: Engineering the Surface Structure of Pd-Based Alloy Catalysts by Design. *ACS Catal.* **2019**, *9*, 781–790. [CrossRef]

Review

From Homogeneous to Heterogenized Molecular Catalysts for H₂ Production by Formic Acid Dehydrogenation: Mechanistic Aspects, Role of Additives, and Co-Catalysts

Panagiota Stathi [1], Maria Solakidou [2], Maria Louloudi [2,*] and Yiannis Deligiannakis [1,*]

[1] Laboratory of Physical Chemistry of Materials & Environment, Department of Physics, University of Ioannina, 45110 Ioannina, Greece; pstathi@cc.uoi.gr
[2] Laboratory of Inorganic Chemistry, Department of Chemistry, University of Ioannina, 45110 Ioannina, Greece; maria-sol@windowslive.com
* Correspondence: mlouloud@uoi.gr (M.L.); ideligia@uoi.gr (Y.D.); Tel.: +30-6937433774 (Y.D.)

Received: 26 December 2019; Accepted: 29 January 2020; Published: 7 February 2020

Abstract: H₂ production via dehydrogenation of formic acid (HCOOH, FA), sodium formate (HCOONa, SF), or their mixtures, at near-ambient conditions, T < 100 °C, P = 1 bar, is intensively pursued, in the context of the most economically and environmentally eligible technologies. Herein we discuss molecular catalysts (ML), consisting of a metal center (M, e.g., Ru, Ir, Fe, Co) and an appropriate ligand (L), which exemplify highly efficient Turnover Numbers (TONs) and Turnover Frequencies (TOFs) in H₂ production from FA/SF. Typically, many of these ML catalysts require the presence of a cofactor that promotes their optimal cycling. Thus, we distinguish the concept of such cofactors in additives vs. co-catalysts: When used at high concentrations, that is stoichiometric amounts vs. the substrate (HCOONa, SF), the cofactors are sacrificial additives. In contrast, co-catalysts are used at much lower concentrations, that is at stoichiometric amount vs. the catalyst. The first part of the present review article discusses the mechanistic key steps and key controversies in the literature, taking into account theoretical modeling data. Then, in the second part, the role of additives and co-catalysts as well as the role of the solvent and the eventual inhibitory role of H₂O are discussed in connection to the main mechanistic steps. For completeness, photons used as activators of ML catalysts are also discussed in the context of co-catalysts. In the third part, we discuss examples of promising hybrid nanocatalysts, consisting of a molecular catalyst ML attached on the surface of a nanoparticle. In the same context, we discuss nanoparticulate co-catalysts and hybrid co-catalysts, consisting of catalyst attached on the surface of a nanoparticle, and their role in the performance of molecular catalysts ML.

Keywords: formic; formate; hybrid; functionalization; hydrogen; co-catalyst; additive; amine; molecular catalyst; nanocatalyst; nano co-catalyst

---

1. Introduction

The constantly increasing energy demand has more than doubled within the past 30 years, thus the present energy demand (~15 TW) (TW = 10^{12} watts) is estimated to approach 30 TW at 2050 [1]. Today, the global energy consumption is covered mainly by fossil fuels (81%), 19% of which are used for transportation [2]. This wide use of fossil fuels—and the ensuing CO_2 emission—constitute one of the major environmental problems [3]. In the quest for clean and renewable technologies, hydrogen is considered as one of the few long-term sustainable clean energy carriers, since, when used in fuel-cells, water is the only by-product [4]. H₂ can be used as an energy carrier because its energy density (143 kJ kg^{-1}) is ~2.6 higher vs. gasoline [5]. It is noticed that hydrogen should be viewed as an energy carrier—rather than a fuel—since a primary energy source is necessary for H_2 generation [6]. In this context, a 'cyclic-economy' concept for 100% renewable hydrogen production can be divided in two main processes: *(i)* CO_2 hydrogenation to produce hydrocarbon fuels and *(ii)* hydrogen generation via dehydrogenation of a hydrocarbon substrate. Currently, 96% of hydrogen generation is based on fossil fuels i.e., 48% from natural gas, 30% from refinery and chemical off-gases, and 18% from coal [5].

In nature, transition metal-based hydrogenases efficiently catalyze H_2 production [7], thus nature may provide inspiration to design new efficient energy technologies. Among these, photoelectrochemical and photocatalytic are the most promising technologies capable of providing clean energy without pollutants and by-products [8]. On the other hand, catalytic H_2 production from organic molecules is a promising approach. Ammonia [9], methanol, ethanol [10], liquid hydrocarbons, and water [11] can be used as substrates. The chemical reactions may involve decomposition, steam reforming, partial oxidation, electrolysis, and gasification [12]. Among these C1 sources of H_2, formic acid (HCOOH, FA) is especially attractive [13,14] since it is a major by-product of biomass processing (formally an adduct of H_2 and CO_2), thus it has attracted considerable attention as a suitable liquid source for H_2 and as a potential H_2-storage material e.g., via a CO_2 hydrogenation reaction [15].

FA can be decomposed to H_2 and CO_2 via dehydrogenation reaction (1).

$$HCOOH_{(l)} \rightarrow H_{2(g)} + CO_{2(g)} \ (\Delta G = -32.9 \ kJ/mol) \tag{1}$$

However, dehydration reaction can take place

$$HCOOH_{(l)} \rightarrow CO_{(g)} + H_2O_{(l)} \ (\Delta G = -12.4 \ kJ/mol) \tag{2}$$

The CO produced in reaction (2) is a fatal poison for catalysts and fuel cells, thus a key prerequisite for the design of any FA-to-H_2 catalyst is to suppress reaction (2). So far, all literary data indicate that the CO-path has low yield in molecular catalysts operating at near ambient condition [16]. Recently, Beller et al. [16] exemplified a unique Pd-L complex able to produce CO at high selectivity. On the other hand, nanoparticulate catalysts can switch between the CO_2 and CO paths depending on the reaction-conditions detail. For example, the *cis*- vs. *trans*-coordination of HCOOH on the particle may trigger one or the other paths [16–18]. Herein we focus on molecular catalysts therefore the issue of CO-path is not further discussed.

Currently, technologies for catalytic transformations of FA to H_2/CO_2 may be classified according *(i)* to the metal-catalyst used i.e., noble metals vs. non-noble metals, and *(ii)* the operational conditions employed i.e., solvent type, temperature, pressure.

The use of noble metal catalysts (i.e., for example Pt, Rh, Pd) [19] has historically pioneered the field of FA dehydrogenation, achieving high performances in terms of TONs, and selectivity. Pertinent cases are discussed in Section 2.1 hereafter. Despite the good performance of noble-metal catalysts, cost and low-abundance issues entailed the need for development of efficient non-noble metal catalysts, which has been intensively pursued during the last decades. Hereafter, in Section 2.2, representative cases for non-noble metal catalysts are discussed in a timeline context. The two classes of catalysts, noble/non-noble metal, are considered within a common mechanistic context, discussed in Sections 2.3 and 2.4. The common mechanistic base of metal complexes and metal–particles are also considered herein. The role of bases and the conceptual difference of the 'additive' (homogeneous) vs. a 'co-catalyst' (heterogeneous) is discussed in Section 2.5.

Regarding the operational catalytic conditions, a key challenge is to achieve high H_2 production rates from FA, at near-ambient conditions i.e., P~1 bar and T < 90 °C, in green solvents, ideally water.

So far, catalytic H_2 production under these constrains, has been shown to be achievable by several types of homogeneous metal complex catalysts [19–21] or noble metal nanoparticles (NPs) [20]. However, the inherent limitations in separation and reusability of the homogeneous catalysts led researchers to explore the possibility to develop heterogeneous catalytic systems.

The catalytic H_2 generation from homogeneous metal complexes has been reviewed by Wills et al. [10] with focus on the type of substrate, by Singh et al. [21] with focus on FA as energy carrier, and recently by Sordakis et al. [19], Onishi et al. [22], and Filonenko et al. [23], comprehensively reviewing the type of metal complex, substrate used, and the reaction conditions and the reversibility of the procedure. Recently, Inglesias and Oro reviewed the homogenous FA dehydrogenation [24]. In our recent works, we have shown that attaching the co-catalyst on a solid matrix i.e., such as SiO_2-particles, provides definitive advantages, boosting H_2 production [25]. Herein, the concept of heterogenized co-catalysts, i.e., derived using covalently attached organic base functionalities, aiming to replace the huge amounts of liquid base typically required so far in this process, is exemplified. From the experimental point of view, the use of heterogenized catalyst or heterogenized co-catalyst [26] allowed a discrete study and understanding of the catalyst/co-catalyst synergy.

The present review focuses on the role of additives and co-catalysts' mechanistic aspects and experimental limitations in catalytic H_2 generation from FA/SF by metal complexes in homogeneous and heterogeneous systems. Herein the term *'heterogeneous'* is used to describe two cases: (*a*) The metal complex is attached on a solid matrix and the co-catalyst is in the homogeneous phase, or (*b*) the co-catalyst is attached on a solid matrix and the metal complex is in the homogeneous phase. Herein, a unified mechanistic reaction view concerning the role of hydride as key factor is discussed in the context of experimental limitations and the intriguing role of co-catalysts.

The catalytic H_2 generation by noble metal nanoparticles is a vast topic [26,27]. Noble-metal NPs are highlighted herein, however only to the extent where it pertains to a comprehension of the mechanistic aspects and the common basis of metal-complexes and metal-nanoparticles. An interesting concept i.e., between nanoparticles and the metal complexes, can be considered the case of 'single-atom' catalytic sites on nanoparticles. This is a new emerging approach. Recently, FA dehydrogenation at high temperatures i.e., 120–160 K from Pt-Cu single-atom alloys has been studied by Flytzani-Stephanopoulos et al., reference [18] indicating that single-atom alloys show six times higher reactivity and selectivity compared with the pure Cu. Photocatalytic H_2 production form FA or methanol is comprehensively discussed in the review of Getoff [28]. Herein, photo-assisted catalytic H_2 generation is discussed in the context of use of light photons as a 'co-catalyst', which affects the conformation of the active center without involvement of redox events. This is distinct from the photocatalytic H_2 generation where light photons provide electrons or protons to the reaction process.

2. Type of Metal Complex-Catalysts

Historically, the choice of metal type i.e., noble vs. non-noble was a matter of strife, since noble metals (e.g., Ru, Ir, Rh) were the first to provide encouraging results for H_2 production at near ambient reaction conditions; see Figure 1.

However, the thrust for low-cost/earth-abundant non-noble metal catalysts (such as Fe, Ni) acted as the driving force to develop competitive non-noble metals catalysts for H_2-production. Herein, for the sake of comprehensiveness, we discuss these two families of catalysts in a comparative manner. In Table 1, we provide a compilation of the most representative cases for non-noble and noble metal complexes.

2.1. Noble Metal Catalysts

In 1967, Coffey [29] was the first to report that metal complexes with phosphine ligands can promote dehydrogenation of FA, in acetic acid as solvent. Among the tested complexes, the best results were obtained with an [IrH$_3$(PPh$_3$)$_3$] complex achieving Turnover Frequencies (TOF) up to 8900 h^{-1}. In the next decades, several types of catalytic systems were reported, reaching reasonable activities i.e., Turnover Numbers (TONs) over 500. In 1998, Puddephatt et al. [30] used a binuclear Ru$_2$(μ-CO)(CO)$_4$(μ-DPPM)$_2$ complex, which in acetone, shows TOF = 70 h^{-1} at RT. Despite these promising early results, only in 2008 was a catalytic system with significantly improved activities and stabilities reported by Laurenczy et al. [31]. This was a water-soluble Ru catalytic system

[Ru(H$_2$O)$_6$](tos)$_2$ (*tos* = toluene-4-sulfonate), which—in a continuous H$_2$ production setup—achieved TOFs > 40.000 at T = 120 °C. This catalytic system is the first to be used—so far—for scalable application and according to the authors is currently under development for the first bus using FA as a fuel [31]. A review of hydrogen storage techniques for on-board applications was presented in 2013 by Durbin and Malardier-Jugroot [32]. All these results are summarized in Table 1.

More recently, a series of Rh(III) or Ir(III) complexes with nitrogen-containing ligands has been evaluated as catalysts for FA dehydrogenation [33]. The Iridium Cp*Ir(TMBI)H$_2$O]SO$_4$ was used for FA dehydrogenation in aqueous solutions at near ambient temperatures achieved TOFs = 34,000 h^{-1}. Another Ir complex [Ir(COD)(NP)](tfO), was studied by Williams and co-workers [33], which was able to perform FA dehydrogenation under continuous operation conditions achieving TOFs of 2,160,000 within 2880 hours. More recently, a [Cp*Ir(1,2-diaminocyclohexane)Cl]Cl catalyst achieved TOFs = 3278 h^{-1} at 90 °C while in the same work it was reported that the homologous Rh(III) catalyst Cp*Rh(bis-NHC)Cl]Na shows a significantly lower efficiency i.e., because of the lower stability of the Rh complex [34]. In 2019, Himeda et al. [35] presented a highly stable and effective Ir-based catalytic complex for H$_2$ production via FA dehydrogenation under high pressure i.e., 40 MPa. Among the tested Ir-complexes, the pyridyl-imidazoline ligand showed the higher activity (TOF = 54,700 h^{-1}) and best stability.

The use of amines to improve the efficiency of the catalytic dehydrogenation reaction of FA was reported by Gao et al. [36] in 2000 who used ethanolamine to enhance the H$_2$ production efficiency by a binuclear Ru complex [36]. Then, a detailed study of the co-catalytic effect of amines in catalytic dehydrogenation of FA was presented by Beller et al., who tested several amines [37]. This work showed that the [FA/amine] ratio is of critical importance for enhancement of HCOOH dehydrogenation at near ambient conditions. Another important finding is the possibility of the ligand itself to actively participate in the catalytic activity of the metal complex, as exemplified by Himeda [38]. In this case, a water-soluble Ir-complex was shown to involve the protonation/deprotonation of the OH group of the 4-4'–dihydroxy-2,2'-bipyridine ligand. As a result, a strong pH dependence of the TOFs was documented i.e., TOF = 2500 h^{-1} at pH = 2.5 vs. TOFs = 500 h^{-1} at pH = 4. The effect of pH on the catalytic dehydrogenation of FA was also reported by Fukuzumi et al. [39].

Overall, the aforementioned works reveal that the efficiency of H$_2$ production from FA by noble-metal catalysts can be strongly modulated by proton-controlling mechanisms. The mechanism of amine can be two-fold.

(i) in *non-aqueous solvents*, amines can be used to boost the FA deprotonation HCOOH→HCOO$^-$,
(ii) in *aqueous solutions*, the amines can have an effect on the HCOOH→HCOO– equilibrium, but also on the ligand itself. Thus, in aqueous solutions, the experimental observation is that pH can control gate the H$_2$ production rate via gating of rate-limiting protonation/deprotonation events.

2.2. Non-Noble Metal Catalysts

So far, as exemplified in Table 1, due to their premium efficiency, noble-metal catalysts have paved the way for catalytic H$_2$ production at near-ambient T, P. These works allowed the scientific community to understand the basic mechanistic limits and prerequisites for the design and utilization of molecular metal catalysts. However, the limited availability and high-cost of the noble metals urge researchers to develop and utilize catalysts based on low-cost, earth-abundant metals. Catalytic dehydrogenation reactions by non-noble metals have been reviewed by Filolenko et al. in their recent review [40] where they present a thorough discussion in catalytic dehydrogenation of amines and alcohols by 3D transition metal complexes.

Since 2010, Beller and co-workers [41] have presented the first Fe-based catalyst i.e., a Fe(CO)$_{12}$ complex, bearing nitrogen and phosphine ligands, with a considerable H$_2$ production efficiency. More precisely, several types of phosphine ligands were evaluated and the inexpensive commercial PPh$_3$ = tris[(2-diphenylphosphino)ethyl] gives the best results [41]. For example, among the tested systems, (Fe/PPh$_3$/TPY) (TPY = 2,2:6,2–tetrpyridine) was the most active (see Table 1) showing TON = 1266, at 60 °C in DMF [41]. Notice that this catalytic system required activation by visible light irradiation with a 300 W xenon lamp [41]. A catalytic mechanism proposed by the authors [41] assumes that light is necessary both for the formation of the active species, as well as the progression of the catalysis, however this is not a photocatalytic system.

Later on, in 2014, Ravasio and co-workers [42] reported that another transition metal i.e., copper, can provide catalytic complexes, with rather simple ligands, that in FA/amine mixture achieved TOF = 25 [1] and TON = 500. More recently, Beller and Laurentzy [43] achieved a breakthrough in this field, developing a highly active Fe-catalyst comprising of Fe(BF4)$_2$ as Fe precursor and PPh3 = tris[(2-diphenylphosphino)ethyl] phosphine as ligand. This Fe/PPh$_3$ catalyst in propylene carbonate solvent is active under ambient conditions with no need of additional bases or light, affording TOF = 9425 h^{-1} at 80 °C [43]. In 2013, Milstein et al. [44] reported the first [Fe/pincer] complex for FA dehydrogenation. The best results were obtained in THF solvent, with the addition of a base [44]. Among the parameters tested was the base concentration, which was found to be crucial i.e., as in the case for noble-metal catalysts, the amount of base required was high: For example, with use of a ratio of FA/NEt3 = 2/1 this system achieved TOF = 500 h^{-1} at 40 °C [44]. A rhenium-based pincer-type complex was synthesized and evaluated for FA dehydrogenation by the same group [45]. The results showed that the H$_2$ production starts above 175 °C [45]. In 2014, groups of Hazari and Schneider demonstrated another Fe-pincer complex with high activity (TOF = 999 h^{-1}) (see Table 1) and stability in the catalytic hydrogen production [46]. More recently, Kirchner and Gonsalvi presented new Fe-pincer catalysts [47] where the effect of additives was studied in detail. In this work, it was shown that amines promote H$_2$ production by dehydrogenation of FA [47].

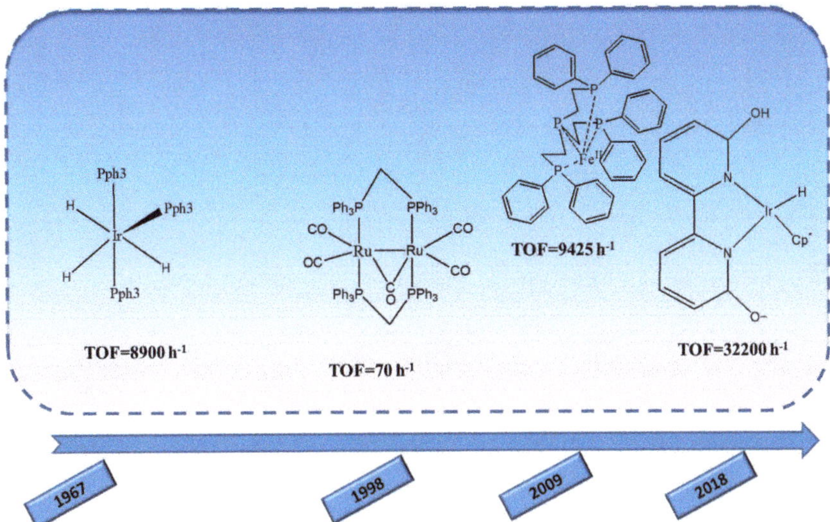

Figure 1. Historical timeline of Turnover Frequencies achieved by representative metal complexes used for formic acid (FA) dehydrogenation, "1967" refers to the work of Coffey [29], "1998" refers to the work of Puddephatt [30], "2009" refers to the work of Beller [43], "2019 refers to the work of Himeda [35].

As shown tin Table 1, a key observation is that in systems where an additive was used, in all reported cases, a *near stoichiometric* amount of amine vs. the substrate is needed. This shows that the additive e.g., an amine in many cases, is sacrificed during the catalytic process. This poses the next challenge i.e., to develop highly efficient processes where either no-additive is required, or the additive can be used in minimal "co-catalytic" amounts. A good premise in this direction provides the examples of the Fe/PPh$_3$ catalyst in propylene carbonate [43], and the Fe-pincer complex [46].

Another challenge is the possibility to develop catalysts able to perform FA-dehydrogenation as well as CO_2-hydrogenation. To this front, recently, Hazari and co-workers [48] reported three novel pincer Fe-isonitrile catalysts, which were active for FA dehydrogenation as well as CO_2 hydrogenation, and achieved TOF = 2100 h^{-1} for FA dehydrogenation.

These pioneering works provide encouraging data that non-noble-metal based catalysts can be tuned to perform FA dehydrogenation as well as also CO_2 hydrogenation. The term "tuned" implies that a detailed understanding of the mechanistic aspects and limiting factors have to be understood and elaborated experimentally in order to proceed to this front. In the following, we discuss the mechanistic landscape aiming to provide a unified view.

2.3. Outline of the Catalytic Mechanisms

All existing literature data show that the molecular mechanism of FA dehydrogenation does *not* involve metal redox reactions (the metal oxidation state stays stable during the catalytic cycle) [16,43,47,49] This is corroborated by spectroscopic studies i.e., EPR [50], NMR, and Uv-Vis [51] data, as well as by theoretical quantum chemical modelling studied.

Cycle-I operates with involvement of the initial $HCOO^-$ anion plus one formic acid (HCOOH) molecule, while Cycle-II with involvement of the initial HCOO– anion plus a second formate anion ($HCOO^-$); both cycles involve β-hydride transfer from substrate (FA or HCOO– anion) to metal.

Accordingly, as we present in Figure 2, the catalytic mechanism can be operationally divided in three sequential steps:

(i) FA activation i.e., formation of a $HCOO^-$ anion. This can be achieved via ionization of HCOONa in aqueous solvent or by deprotonation of formic acid HCOOH in aprotic solvents, see Figure 2.

(ii) Catalyst activation: This might involve coordination of one HCOO– on the (LM) complex, or formation of the (LM-H) hydride via LnM/H_2 interaction at high H_2 pressure.

(iii) Catalytic H_2 production. This critical step can be accomplished with two alternative routes, exemplified as Cycle-I and Cycle-II in Figure 2.

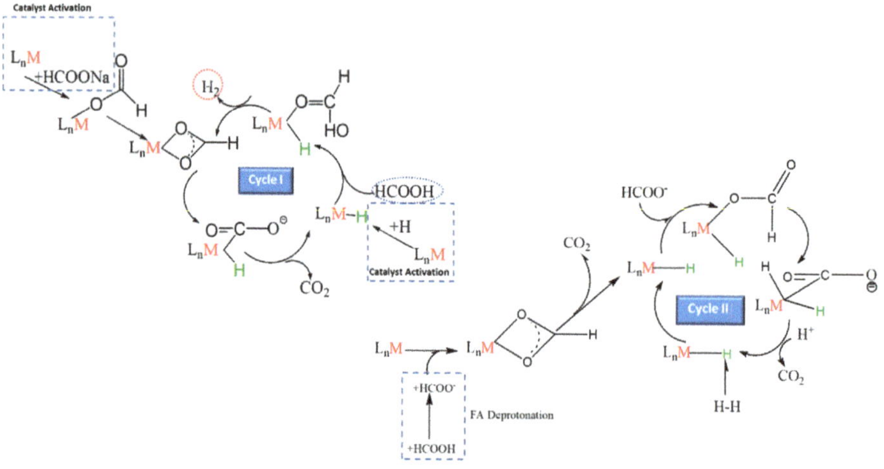

Figure 2. General catalytic landscape of FA dehydrogenation by metal-complex catalyst in solution.

Cycle-I operates with involvement of the initial $HCOO^-$ anion plus one formic acid (HCOOH) molecule, while Cycle-II with involvement of the initial $HCOO^-$ anion plus a second formate anion ($HCOO^-$); both cycles involve β-hydride transfer from substrate (FA or $HCOO^-$ anion) to metal.

Cycle I: This cycle initiates with the direct reaction of a monohydride (H-ML) complex with a formic acid molecule towards formation of the (H-ML)-HCOOH transient state. Subsequently, a—fast—hydride protonation step takes place [51], refernece [52] with subsequent H_2 formation and liberation. A chelate coordination of the formate anion to metal, followed by hydride generation and CO_2 release, are proposed [52,53] to be the steps that complete the cycle. However, this involves a rather complex sequence of events, as exemplified by the theoretical work [53] depicted in Figure 3.

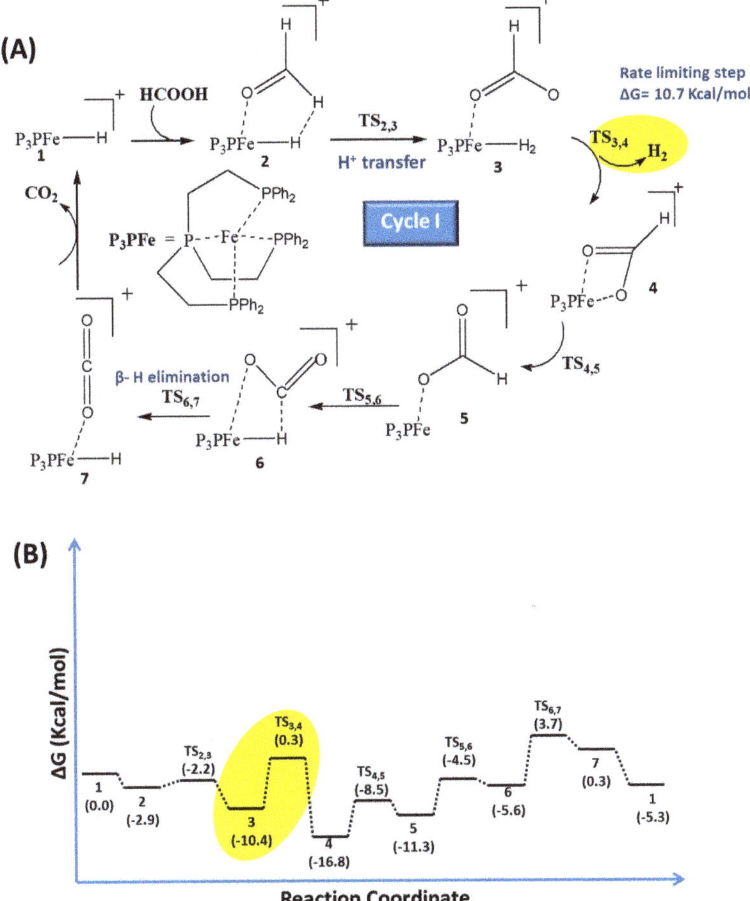

Figure 3. Catalytic (Cycle-I) for the [FePPh3/HCOOH] system (**A**) and free energy profile (**B**). Figure adapted from Reference [52]. In Cycle-I, the reaction is rate-limited in transition state 3,4 by the H-H release from the Fe-center.

At this point we have to notice that, in some cases, for the operation of Cycle-I, the crucial monohydride (H-ML) complex can be formed via pre-incubation of the catalyst with H_2 at high pressures as shown by Fellay et al. [49]. Actually, this strategy may be considered as a 'catalyst activation' step; see Figure 2. In the absence of such catalyst pre-incubation, ready coordination of spare amounts of formate anion can initially boost the monohydride (H-ML) formation i.e., functioning as another type of catalyst activation; see Figure 2. A more comprehensive elaboration on Cycle-I is discussed hereafter, based on theoretical quantum calculations.

Cycle II: Starting with the activation/deprotonation of FA, the formate anion promotes the formation of the monohydride (H-ML) complex. The formate anion could be inserted to the catalytic cycle via its coordination to metal monohydride (H-ML) species towards a (metal)-dihydride formation (see

Figure 2) via a β-hydride elimination step, and a transiently coordinated O=C=O molecule. Then, liberation of CO_2 and subsequent coordination of a dihydrogen, H-H, molecule generated after the synergy of a H^+ results in gaseous H_2 release.

   A case study: In 2013, Yang [52] presented a detailed theoretical study for the system [Fe/(CH$_2$CH$_2$PPh$_2$)$_3$]. Despite the fact that the dielectric/solvent properties are not taken into account in these calculations, these theoretical data provide useful insights into the mechanism. Detailed free energy profiles were obtained for the two competitive cycles and confirmed that the β-hydride elimination is the rate determining step for the catalytic Cycle-I, which operates with FA and total free energy cost is 20.5 kcal/mol (i.e., 85.8 kJ/mol). The total energetic cost for Cycle-II, which uses formate anion as substrate, was calculated to be 22.8 kcal/mol or 95.4 kJ/mol, see Figure 4. The theoretical data presented by this author [52] indicate that Cycle-II is slightly less favorable than Cycle-I. However, one should take into account that within the limitations of the DFT calculations, the preference of Cycle-I or Cycle-II might rely on small energy differences, which means that changes in experimental conditions (different metal, ligands, solvents) may favor either Cycle-I or Cycle-II.

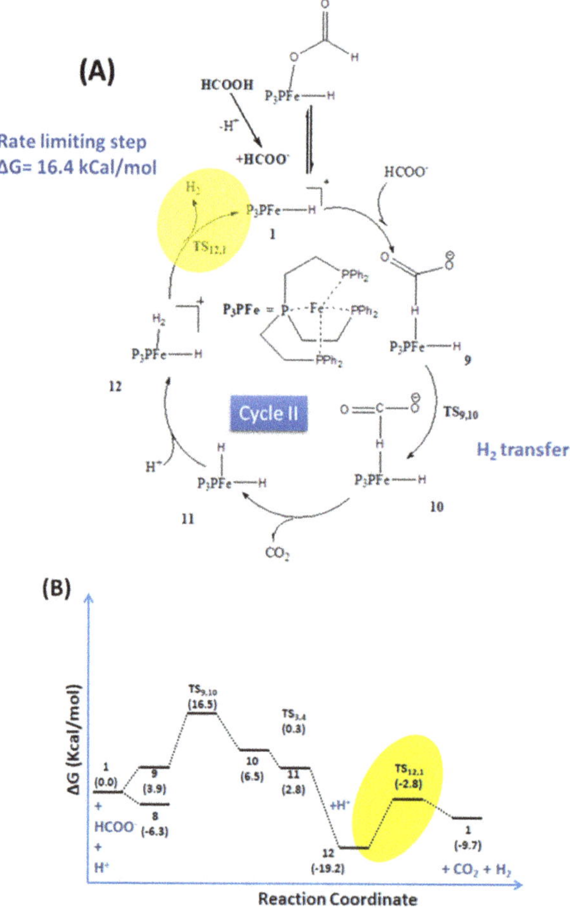

Figure 4. Catalytic Cycle-II for the [FePPh3/HCOOH] system (**A**) and free energy profile (**B**). Figure Adapted from Reference [52]. In Cycle-II, the reaction is rate-limited by the transition state 12,1 regarding H-H release from the Fe-center.

As shown in Figure 3 for cycle-I, bidentate coordination of a HCOOH molecule is strongly favored i.e., a stabilization by $\Delta G = -16.8$ kcal/mol (-70.3 kJ/mol) (intermediate 4).

In contrast, the transition state 3,4 i.e., release of H_2 from intermediate 3, is unfavored by a barrier $\Delta G = +0.3 - (-10.4) = +10.7$ kcal/mol. Thus, according to these calculations, the release of the coordinated dihydrogen is the rate-limiting step in Cycle-I.

In Cycle-II, the intramolecular H transfer step (transition state 9,10) is unfavored by a barrier $\Delta G = +16.5 - 3.9 = +13.4$ kcal/mol. Here, the rate limiting step is transition state 12,1, i.e., the release of the coordinated dihydrogen faces an energy barrier $\Delta G = -2.8 - (-19.2) = +16.4$ kcal/mol.

According to these calculations, both Cycle-I and -II are rate limited by the H-H release from the Fe-center. Cycle-II is more energy demanding i.e., $\Delta G = +16.4$ kcal/mol $= 68.6$ kJ/mol vs. 56.0 kJ/mole in Cycle-I.

Thus, it is expected that the H_2 production by this system will proceed mainly via Cycle-I [52]. We should underline that, as noticed by Yang [52], the energy cost of the FA deprotonation was not taken into account in these calculation because of the uncertainty of the dissociation energy of FA, which may be affected by the ligand, the counter ions in the solution, and the solvent [52]. This highlights the crucial role of the solvent to be selected in conjunction with the other limitations imposed by the catalytic system. Analogous theoretical calculations for a similar Fe-complex were performed by Ahlquist and co-workers [53].

According to this theoretical work [53], the results show that Cycle-II is more favorable i.e., total energy cost = 86.2 kJ/mol for Cycle-II vs. 128 kJ/mol for Cycle-I; see Figure 5.

The discrepancies between reference [52] and [53] indicate that the small energy differences of the two cycles and the limited capacity of the computational approaches to incorporate all factors i.e., solvent effects, ionization of FA, etc., should be considered in future refinement of theoretical quantum calculations. Also, currently, the mechanistic discussion is mainly based on theoretical results, therefore a critical assessment of the boundaries of DFT-methodology is mandatory. In this context, apart from the sensitivity of the computational complications to solvent effects, traces of H_2O, and co-catalysts, we notice that the ground-state spin multiplicity for the proposed hydride catalyst species in Yangs' work [52] is different than that used in Ahlquist's work [53].

Despite the methodological limitations of the theoretical analysis, experimental work [48] provides evidence that the energy barriers in Cycle-I and Cycle-II are influenced by the experimental parameters (type of complex, solvent, traces of H_2O, FA deprotonation, presence of co-catalysts), which might decisively affect the contribution of each cycle to the overall H_2 production rates. In this context, we have experimentally demonstrated that silica nanoparticles can act as co-catalyst to the Fe/PPh$_3$ system [54] (see Figure 6) suppressing the energy barrier of the reaction from 77 kJ/mol to 36 kJ/mol [54].

This effect is attributed to the deprotonation of the FA (FA activation) before insertion in the catalytic cycle via bidentate coordination facilitating hydride formation [54]. This step drives the overall reaction to operate via Cycle-II. Interestingly, the two different cycles we have recently experimentally evidenced are for FA dehydrogenation by Pd-nanoparticles [55]. The two proposed cycles differ in the form of the FA used as substrate—HCOOH vs. HCOO$^-$ (see Figure 7) [55]. Accordingly, as shown in Reference [55], excess of HCOO$^-$ anion is required to achieve enhanced H_2 production rates, indicating that Cycle-II is the thermodynamically dominant. Importantly, it has been shown [55] that deprotonation of HCOOH towards HCOO$^-$ can be achieved by adjacent/grafted additives i.e., gallic acid, and this boosts the H_2 production.

A different possible catalytic cycle for FA dehydrogenation was proposed in a pioneering work of Himeda et al. [56]; see Figure 8. In this rather non-classical cycle, the rate-limiting step is the β-hydride elimination from the (metal-HCOO$^-$) complex by the hydroxyl groups of the ligand. The β-hydride elimination takes place via intermolecular rearrangement to an H-bond between a formate and the ligand [56] (Figure 8). As expected, the pH of the solution plays a crucial role in the operation of the cycle and the rate limiting step of the mechanism. At pH = 1, TONs were 10,000 while at pH = 5,

TONs were very low i.e., 200; as noticed by the authors, at higher pH values, no gas evolution was observed [56].

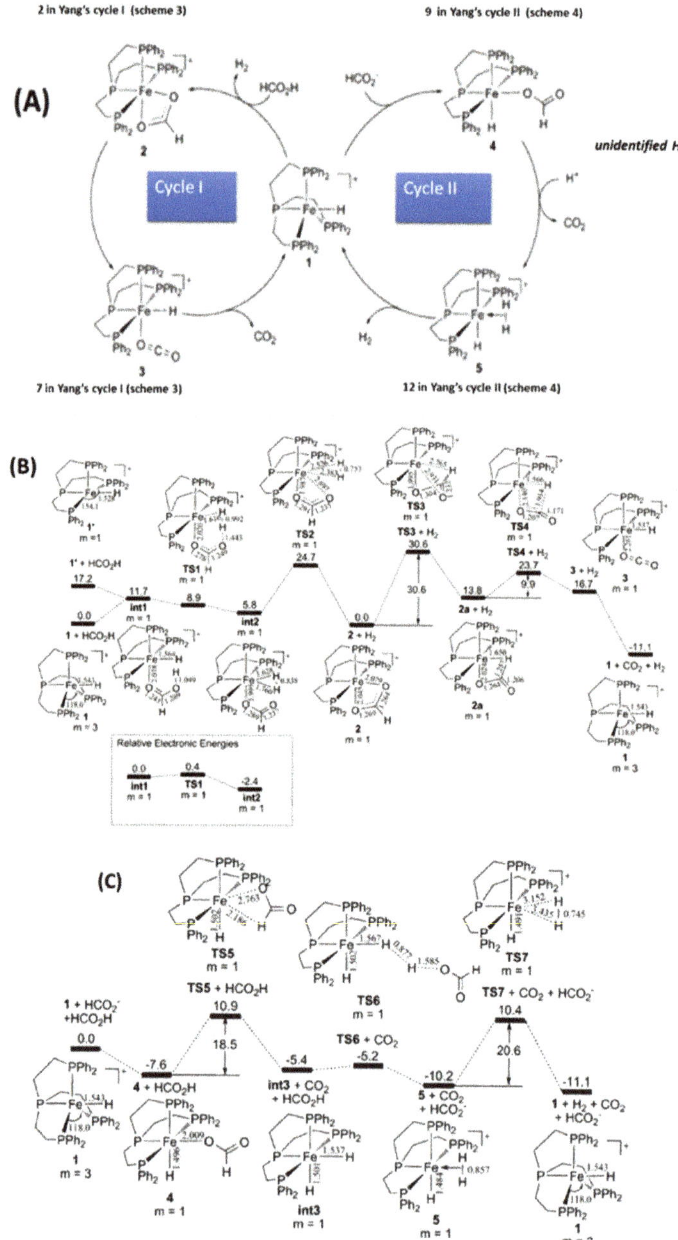

Figure 5. Catalytic Cycle-I and Cycle-II (**A**) and the energetic profiles as calculated by Ahlquist et al. [53]; (**B**) the transient intermediates for cycle-I and (**C**) the transient intermediates for cycle II. Figure adapted from Reference [53].

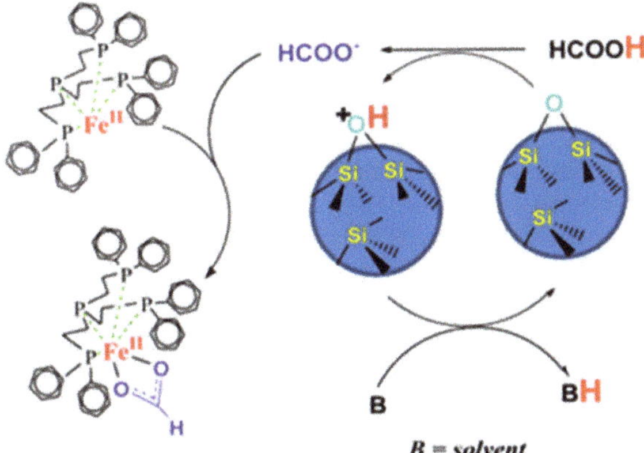

Figure 6. Co-catalytic mechanism of SiO$_2$ nanoparticles. Figure adapted from Reference [54]. Silica nanoparticles can act as co-catalyst to the Fe/PPh$_3$ system by promotion of the HCOOH deprotonation.

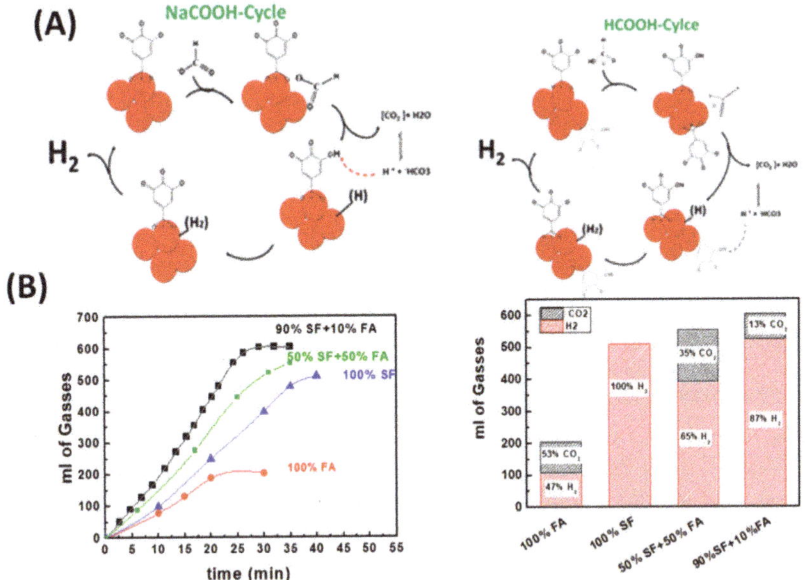

Figure 7. (**A**) Proposed catalytic cycles for Pd@SiO$_2$-GA nanoparticles (NPs); (**B**) catalytic results obtained from Pd@SiO$_2$-GA system. Figure adapted from Reference [55].

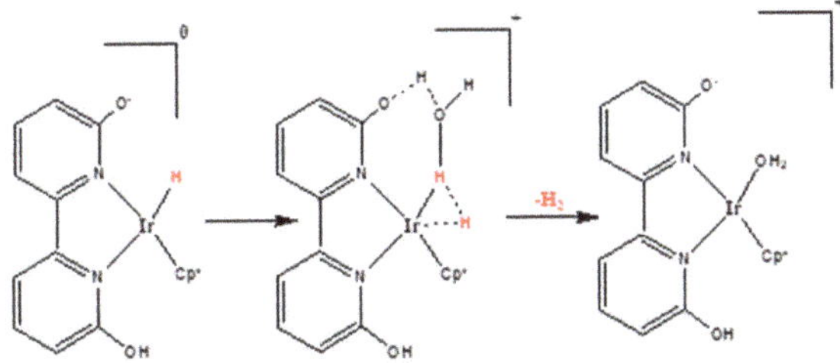

Figure 8. Effect of the pH-tunable ligand in the FA dehydrogenation reaction. Figure adapted from Reference [56].

In the same context, participation of the ligand in the catalytic mechanism has been proposed by Manca et al. [57] in the case of Ru complexes. They theoretically studied two different Ru complexes: a [Ru(κ4NP3)Cl$_2$] operating via Cycle-I and Cycle-II [57]. On the other hand, in the case of [Ru(κ_3-triphos)(MeCN)] complex, participation of the ligand in formation of the active species is proposed [57].

The Unidentified H$^+$: When peering into the proposed catalytic Cycle-II, we notice that an extra proton H$^+$ needs to enter the cycle i.e., see step- 11$\rightarrow$12 in Figure 4 (Yang [52]), to 4$\rightarrow$5 in Figure 5. This was also invoked in our proposed mechanism the SiO$_2$ nanoparticles (see Figure 6) [55]. Despite the fact that the availability of this H$^+$ is crucial, its origin is not clear. While in aqueous media, H$^+$ are abundant, in aprotic media i.e., such as propylene carbonate, availability of this proton should be considered with care. So far, the origin of this proton has not been proven is none of the published works, including ours [25,54]. Traces of H$_2$O present in the solvent can be invoked, however since these H$^+$ participate stoichiometrically for each one HCOO$^-$ converted to H$_2$, the H$_2$O content should be able to provide H$^+$ at *high concentrations* i.e., stoichiometric vs. the HCOO$^-$ substrate. In a typical experimental setup e.g., consider the case where 0.026 mmoles of HCOO$^-$ are added in 5 mL of solvent, the extra proton would require the existence of several mmoles/lt of H$_2$O, which is impossible. Alternatively, "spectator ligands" i.e., ligands not coordinating a metal center can be considered a possible source of these stoichiometric protons. However, a targeted study of this possibility remains to be done.

Can we distinguish Cycle-I or Cycle-II by the [substrate:H$_2$ stoichiometry]; according to Figure 2, the stoichiometry of [HCOOH:H$_2$] in Cycle-I is 1:1, while the stoichiometry [HCOO$^-$:H$_2$] in Cycle-II is 1:1. A key point is that the sacrificial HCOO$^-$ that is consumed in the initial step activation of the catalyst i.e., resulting in the formation of the first stable metal-hydride (see LM-H Figure 2) participates only at the beginning of the cycles I, II. Thus, this initial HCOO$^-$ does *not* count in the total stoichiometry of the produced H$_2$. Otherwise, i.e., if *two* HCOO$^-$ were required for each H$_2$ molecule, then the stoichiometry HCOO$^-$:H$_2$ would be 2:1. This case i.e., a 2:1 stoichiometry of [HCOO$^-$:H$_2$], has not been reported so far. In all reported cases, a ~1:1 stoichiometry is observed i.e., consistent with cycle-I and/or Cycle-II. Thus, the [HCOOH:H$_2$] stoichiometry itself cannot distinguish Cycle-I from Cycle-II.

2.4. Catalyst Activation

2.4.1. Based on the Applied Preparation Method

Based on the applied preparation method, the metal complexes used for catalytic FA dehydrogenation can be classified in two categories: *(i)* Pre-formed (ML) complexes and *(ii)* in situ formed complexes, that is, a metal precursor e.g., metal salt, and the ligand L [19] is mixed in the reaction mixture, immediately before the addition of the substrate [43]. Then, after formation of the complex (ML), the coordination of a formate anion HCOO$^-$ to the metal center is the first step (Figure 2). For this reason, preincubation of the complex with formate ions is required [47]. Alternatively, formation of the active hydride LnM-H species can be promoted by incubation with high-pressure H$_2$ (See Figure 2). Fellay et al. [49] showed that preincubation with H$_2$ at pressure P$_{H2}$ = 750 bar is enough to drive the formation of the LnM-H species.

2.4.2. Light-Assisted Activation (LAA) of Catalyst

Another interesting aspect for the catalytic FA dehydrogenation is the possibility of acceleration of the catalytic reaction by light photons. This is a *"Light-Assisted Activation of the LnM complex"*, (LAA) without redox activity of M. The LAA mechanism can involve one or multiple photons:

One-photon LAA: Linn et al. [58] tested the catalytic performance of the Cr(CO)$_6$ catalyst with a HCOONa salt for H$_2$ production. It was shown [58] that visible light photons can promote the formation of the catalytically active penta-carbonyl Cr(CO)$_5$ complex via a one-photon dissociation of one CO from the Cr(CO)$_6$ (Figure 9).

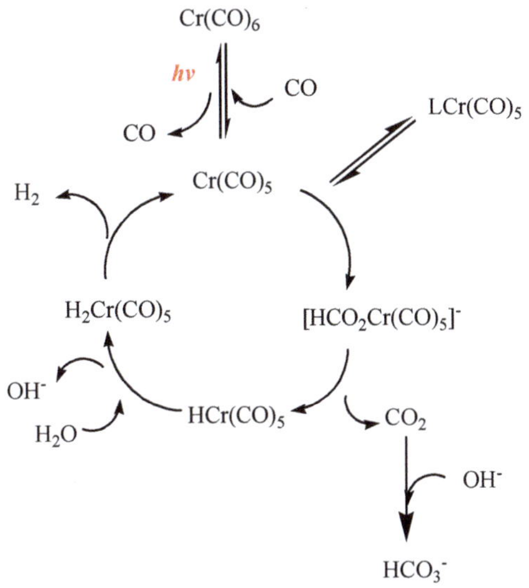

Figure 9. Catalytic cycle for FA dehydrogenation one-photon Light-Assisted Activation (LAA) mechanism. Figure adapted from Reference [58].

Multiple-photon activation Cycles: In early 1990, Onishi [59] tested the photo-assisted efficiency of a hydrido(phosphonite)cobalt(I) [CoH{PPh(OEt)$_2$}$_4$] complex for dehydrogenation of HCOOH. As depicted in Figure 10, the H$_2$ production is associated with the formation of the complex [CoH$_2${PPh(OEt)$_2$})$_4$]$^+$ (HCOO$^-$) from CoHP$_4$. One-photon excitation of this complex leads to the release of dihydrogen and the formation of the formato-complex, [Co(O$_2$CH){PPh(OEt)$_2$}$_4$]. Then, a second photon promotes release of CO$_2$. Thus, two-photon excitation of the system drives the metal catalyst to the CoHP$_4$ state. The overall mechanism is illustrated below in Figure 10.

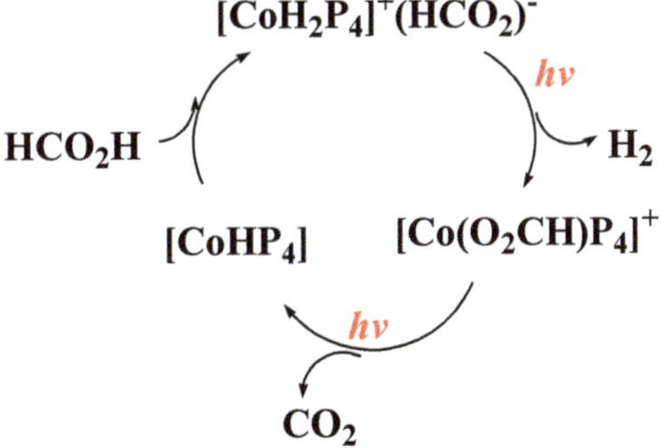

Figure 10. Catalytic cycle for FA dehydrogenation-assisted multiple-photons mechanism. Figure adapted from Reference [59].

In 2009, Beller et al. [60] investigated the effect of light on H$_2$ generation from formic acid, using a catalyst system based on a ruthenium precursor and aryl phosphines [Ru(Paryl$_3$)m)]; see Figure 11. In comparison with the non-photo-assisted system, the gas evolution achieved by photo assistance was higher by ~200%.

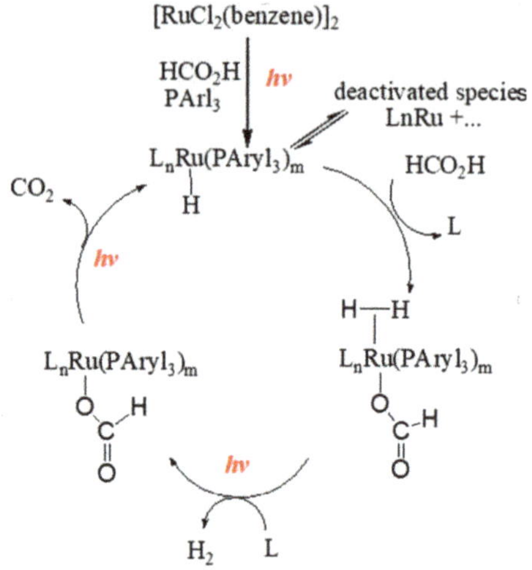

Figure 11. Catalytic cycle for FA dehydrogenation proposed by Beller et al. with assisted multiple-photons mechanism. Figure adapted from Reference [60].

The proposed catalytic mechanism entails involvement of up to four photons of visible irradiation, (excluding UV < 320 nm and IR irradiation, [60]). As shown in Figure 11, the initial [RuCl$_2$(benz)]$_2$ pre-catalyst is activated by one visible light photon, to give an aryl phosphine-ligated Ru-hydride species. Deactivation of this active catalyst is prevented by light. A second photon promotes release of H$_2$ while a third photon promotes dissociation of CO$_2$ from the ligated HCOO$^-$ anion; see Figure 11. Thus, in all the reported cases so far, light photons may be promoters of the *H$_2$-release* from the catalytic intermediate complex. Since, according to the theoretical analysis, this H$_2$-release is expected to be the rate-limiting step, the photons can accelerate the H$_2$ production rate by their targeted intervention at this step.

2.5. Formic Acid Deprotonation

2.5.1. Deprotonation by Homogeneous Molecular Co-catalysts

Bases as co-catalysts: As described earlier, in Figures 2–4, participation of a HCOO$^-$ anion is mandatory in Cycle-II. Moreover, as detailed in Figure 2, the formation of the first metal-hydride (LnM-H) can be established via an initial coordination of one HCOO$^-$ to the LnM complex. Thus, the availability of HCOO$^-$ is expected to be crucial for both Cycle-I and Cycle-II. The continuous dependence of Cycle-II on available HCOO$^-$ anions renders the systems where Cycle-II dominates more sensitive to the activation of formic acid i.e., the deprotonation towards HCOO$^-$ anions. In this context, taking into account the importance of the FA deprotonation for the efficient catalytic performance of the system, several types of additives have been tested. At this point it is didactic to underline that use of HCOONa (Sodium Formate, SF) in small amounts as a basic additive, are beneficial i.e., promote Cycle-II via HCOO$^-$. It should be clarified however that when SF is used as the main substrate (high concentrations), the reaction can produce bicarbonate anions i.e., as catalytic byproduct, that can be poisonous for the dehydrogenation reaction. This has been clearly exemplified in the case of H$_2$ production by [Pd-Galic Acid] nanoparticles (see Figure 7) [55] where the excess of bicarbonate resulted in a layer of bicarbonate crystals deposited on the catalyst surface.

In some cases, the anionic form of the ligand can act as an internal base, thus such systems work under base-free conditions. The key point in these systems is the ionization of the ligand, affected by the pH of the solution. In this context, Fukuzumi proposed pH-tunable catalytic systems.

Another approach for FA activation is the use of organic bases as additives or solvents. In this context, the influence of the organic bases in the catalytic performance of Ru catalysts was studied by Boddien et al. [61]. Several amines, pyridines and ureas were tested, see Figure 12, as additives; the best results were obtained for the tertiary alkyl amines, i.e., like triethyl-amine. We underline that in [61] the [amine/FA] ratios tested were near-stoichiometric i.e., in the range 2/5 to 4/5, thus the amines acting as sacrificial 'additives' rather than co-catalsysts. In general, there is a correlation between amines' basicity and TONs achieved (see Figure 12B). For example, in the same work [61] with [RuCl$_2$(p-cymene)] as catalyst, an [amine]:[FA] ratio equal to [3:4] provided TOF = 38 h^{-1}, while a molar ratio amine: FA = 2:5 gave lower TOF = 14 h^{-1}.

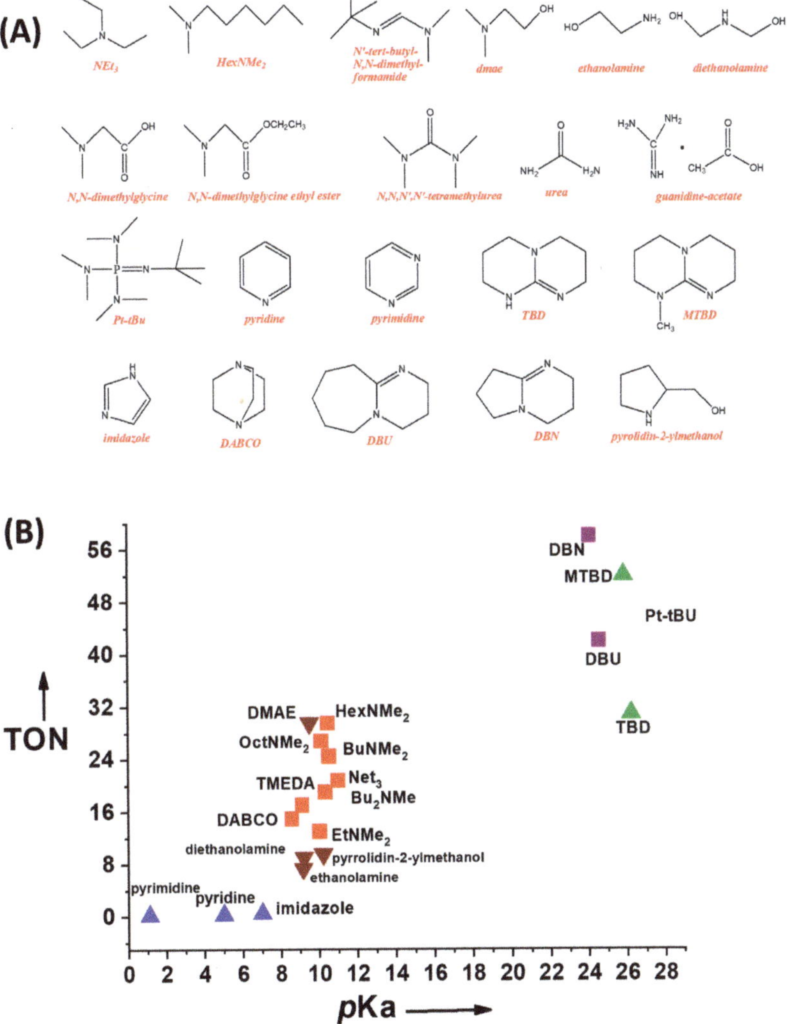

Figure 12. (**A**) Organic molecules used as a co-catalyst in the FA dehydrogenation reaction. (**B**) Correlation between the pKa of the co-catalysts with the Turnover Numbers (TONs) achieved. Figure adapted from Reference [61].

Based on this behavior, it is considered that the amine additives are sacrificed, since almost one amine molecule is required per FA molecule i.e., for every FA molecule to be deprotonated in order to be incorporated in Cycle-II.

The chemical stability of the amine additives is also a factor of concern, especially under the need of continuous operation of the catalytic systems. In continuously operating systems, less volatile amines have to be used to avoid their volatilization. In some cases, the replacement of the volatile triethylamine with the less-volatile N,N-dimethylhexylamine [61] or with the higher-boiling-point dimethyloctyamine improves the catalytic activity of the system [61].

Lewis Acid as co-catalyst: In 2004, Hazari [46] and co-workers presented the synthesis and characterization of a series of Pincer-Fe complexes and their use for FA dehydrogenation, by addition of Lewis acid co-catalysts i.e., instead of a base as co-catalyst. Based on spectroscopic results, the authors proposed a mechanism for FA dehydrogenation [46]. In this mechanism, the pincer-Fe-complex forms a pincer-Fe-HCOO$^-$ adduct using one HCOOH. Subsequent Lewis-acid-assisted decarboxylation of the Fe-complex forms dihydrides with liberation of H_2 and CO_2.

2.5.2. Deprotonation of FA by Heterogeneous Co-catalytic-particles

As exemplified in Section 2.5.1 and references reported herein, amines in the homogeneous phase can act stoichiometrically i.e., as sacrificial agent for deprotonation of one HCOOH to HCOO$^-$. Therefore, high amounts of amines are required to accomplish this task for the entire FA feedstock.

Recently, our group demonstrated an alternative to this requirement, showing that—instead of amines—SiO_2 nanoparticles can enhance H_2 production by a Fe/PPh$_3$/FA catalyst, in propylene carbonate solvent [54]. This study provided the first evidence that the H_2-production boosting effect by SiO_2 -particles was linked to the FA deprotonation e.g., enhanced catalytic H_2 production is promoted by the basic sites of SiO_2 particles. As shown [54], SiO_2 particles initiate deprotonation of HCOOH towards HCOO$^-$ accelerating the formation of the metal-hydride species, thus enhancing the efficiency of Cycle-I. Importantly, we have shown that SiO_2 particles operate in co-catalytic amounts i.e., surface OH-groups at stoichiometric amounts vs. the metal center, SiO2 (OH):M~1:1 [54].

More specifically, the TOFs of the [Fe/PPh$_3$] system have been shown to increase from 1206 h^{-1} to 13,882 h^{-1} after addition of SiO_2. Additionally, by controlling the population of surface groups of SiO_2 i.e., number of Si-OH vs. Si-O-Si, we can control the efficiency of H_2 production; this is because the Si-O-Si bridges promote the FA deprotonation due to their better ability for proton abstraction [54]. In this work, the beneficial effect of the SiO_2 NPs can be optimal at ratio FA/[≡Si-O-Si] = 200/1,

In the same context, SiO_2-NPs modified by organic basic functionalities were shown to provide a significant beneficial effect in the catalytic efficiency of the Fe/phosphine system [62]. More particularly, the use of imidazole-functionalized SiO_2 increased TOFs from 2412 h^{-1} to 16,432 h^{-1}, while the use of NH$_2$-functionalized -SiO_2 increased TOFs from 2412 h^{-1} to 14,942 h^{-1}. In these cases, the best results were achieved with FA/[≡Imid] = 577/1 (56 µmol of imidazole groups over 26 mmol of FA), and FA/[≡NH$_2$] = 500/1(48 µmol of NH$_2$ -groups over 26 mmol of FA).

More recently, heterogenized amino functionalities were used as co-catalyst with a homogeneous Ru/PPh$_3$/FA system [24] see Figure 13. It was found that the heterogeneous H$_2$N@SiO$_2$ co-catalyst (97 µmol of NH$_2$ -groups over 52 mmol of FA) is able to enhance dramatically the H_2 Production by >700% vs. the process without co-catalyst, while the same amount of liquid *n*-propylamine produces six times lower H_2 volume at the same time lapse. Additionally, the heterogenized co-catalyst can be regenerated and re-used. The H$_2$N@SiO$_2$ co-catalyst could be used four times with a < 5% loss of its activity after each reuse [24].

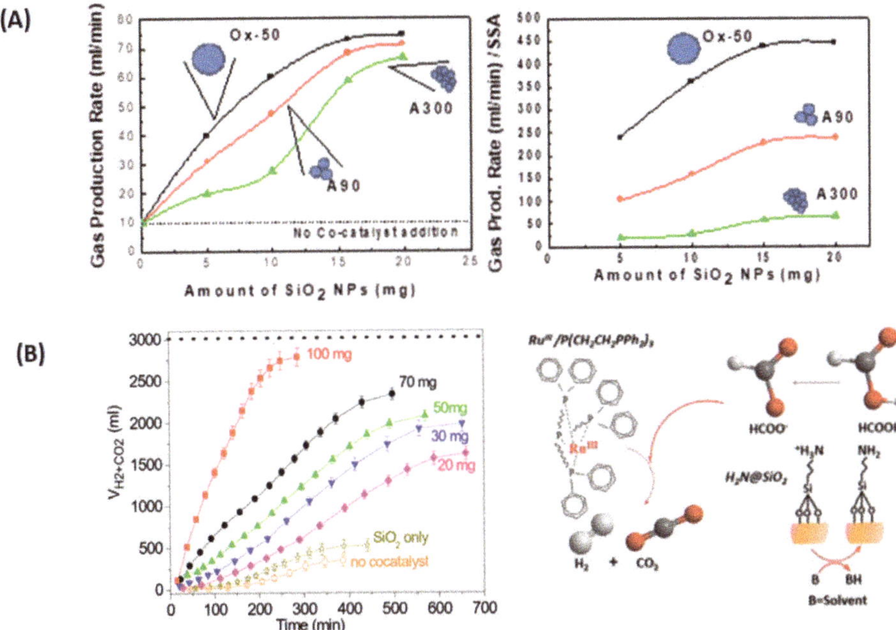

Figure 13. (**A**) Effect of SiO_2 NPs on the catalytic efficiency of Fe/PPh3 system. Figure adapted from Reference [54]. (**B**) Effect of amino functionalized SiO_2 NPs on the catalytic efficiency of Ru/PPh3 system. Figure adapted from Reference [24].

3. Experimental Issues—Limitations

Several parameters should be optimized in tandem for maximum H_2 production via FA dehydrogenation. Here we discuss some experimental issues and limitations pertaining to the catalytic experiments of H_2 production by molecular catalysts at near-ambient conditions.

3.1. Solution-Chemistry Issues

(i) *The choice of solvent*: Several types of solvents have been used so far for near-ambient T and P catalytic FA dehydrogenation reactions [19]. The choice of the solvent is dictated by three parameters: *(i)* Catalyst solubility in the particular solvent, *(ii)* co-catalyst's solubility, in the particular solvent, and *(iii)* solvent polarity. Apart from solubility issues, protic solvents may be more directly implicated in the reaction steps involving H-bonding, deprotonation events. These are detailed in paragraph *(iii)* hereafter, while the eventual poisonous effect of H_2O is discussed in paragraph *(iv)*.

Keep in mind that formic acid can be used as solvent too, thus FA itself imposes the reaction parameters. We have to notice that for noble metal complexes, the use of water as solvent is also possible. Taking into account that some reactions operate better at higher temperatures i.e., 80–120 °C, solvent stability and volatility becomes a crucial parameter. Moreover, a wide variety of investigations have been realized on the effect protic and non-protic solvents on H_2 production, related to the solubility of catalysts. For example, a [Fe-L$_{pincer}$-CO] complex in FA/Net3 solution was tested in protic and non-protic polar solvents [44]. At 40 °C, it was found that the highest catalytic yields were obtained with 1,4-dioxane (TONs = 653), while DMSO, THF, EtOH, CH3CN, and H_2O gave 538, 520, 243, 206, and 83 TONs, respectively [44]. According to the authors, this was attributed to the poor solubility of the complex to more-polar solvents. For a [Fe/PPh3] complex, the best catalytic activity was achieved using propylene carbonate solvent (Ea = 77 kJ/mol) instead of THF (Ea = 82.1 kJ/mol) [44].

(ii) *The influence of reaction temperature*: According to all theoretical as well as the experimental data, H_2 release from HCOOH/HCOO$^-$ is a thermally activated process occurring via high-energy reaction intermediates. Thus, increase of the reaction temperature results in enhancement of the production rate i.e., TOFs. A typical example is the catalyst Fe/PPh$_3$ in propylene carbonate, which shows TOFs = 2018 h^{-1} at 60 °C, and 8136 h^{-1} at 80 °C [36]. It was found that—for the catalysts studied so far in literature—the activation energy barriers, Ea, of these processes are usually in the range of 80 to 40 kJ/mole [36]. On the practical side, such Ea values require operating at temperatures near to the limit of volatility of the solvent, in order to obtain the optimal catalytic rates. This entails that lowering the E_a would allow lower operation temperatures to be applied.

(iii) *Role of pH*: Noble and non-noble metal complexes in aqueous solution may show a pH-dependent catalytic activity i.e., at the pH region that favors FA deprotonation with an optimum efficiency in slightly acidic pH values from 4-6. Transition metal complexes like Ir, Ru, and Rh present a better catalytic activity in the presence of a base. For example, Fukuzumi et al. [39] using a Ru-Ir heterometalic complex [IrIII(Cp*)(H$_2$O)(bpm)RuII(bpy)$_2$](SO$_4$)$_2$] in H$_2$O, demonstrated clearly a characteristic pH-dependence of TOFs, attaining TOF = 450 h^{-1} at optimum pH = 3.8. This pH value is correlated with the pKa of HCOOH deprotonation. Thus, solution pH modulates FA deprotonation making HCOO$^-$ anion the appropriate feed agent for Cycle-II in Figure 2. At more acidic pH < 3, the achieved catalytic activity was lower due to poor possibility of formate anion species, while at pH > 5, [H$_3$O$^+$] concentration is practically low and thus unable to protonate effectively the metal hydride towards H$_2$ formation [39]. In a similar way, the pH-dependent H$_2$ production by an Ir-complex (cis-mer-[IrH$_2$Cl(mtppms)$_3$]) in the presence of HCOOH/HCOONa revealed a sharp maximum of the rate at pH = 3.75, (coinciding with the pKa of formic acid) showing that both H$^+$ or HCOOH and HCOO$^-$ species play important role in the reaction mechanism [63].

(iv) *The poisonous effect of H_2O and Inorganic Anions*: In aprotic solvents, the presence of trace amounts of water can be a limiting factor. For example, we have found [23] that the performance of [Fe/PPh$_3$/propylene carbonate] system can be severely inhibited when H$_2$O was present over a certain limiting amount i.e., 6% [23]. Notice that commercial formic acid typically contains at least 2–3% of water to sustain its liquid phase. Analogous poisonous results were observed by Mellman et al. [51] when Cl$^-$ anions were present in the catalytic mixture. The so-obtained decrease in the catalytic activity can be attributed to the formation of Fe-chloride species that inhibits the Fe-hydride formation, which, as we have discussed herein (Figure 2) is extremely crucial for both Cycle-I and Cycle-II. *This entails that any atom/molecule coordinating irreversibly the metal center—apart from the hydride—will act as inhibitor of the catalytic H_2 production by the metal complexes.*

3.2. Hydrogen Production Under Continuous Operation

Many systems have been described in the literature as able to produce hydrogen via FA dehydrogenation at satisfactory TONs and TOFs. However, for commercial applications, continuous operation of the catalytic system is required. The simplest principle of a continuous operating systems is that the substrate (HCOOH) is provided continuously by an on-line pump; see Figure 14. The crucial parameter in the continuous operation systems is the long-term stability/durability of the catalytic system under the reaction conditions.

Moreover, a robust co-catalyst is required. So far, there are only few examples of catalytic systems working satisfactory under continuous operation. The first catalytic system of continuous hydrogen production via FA dehydrogenation was reported by Laurenczy and co-workers in 2008 [31]. This concept was later scaled up and drove to the development of the FA-fuel vehicle. In 2009, Beller and co-workers reported a Ru-based complex [RuCl$_2$(C$_6$H$_6$)]$_2$/DPPE [64] evaluated under continuous operation conditions and after 11 days of continuous operation achieved TONs 260,000. Wills [65] presented a [RuCl$_2$(DMSO)$_4$] catalyst that was evaluated under continuous operation conditions showing a remarkable TOF = 18,000 h^{-1}. Further improvement of the Wills' system was achieved by replacement of the volatile NEt$_3$ with more stable aliphatic and aromatic amines [66]. Another example for continuous hydrogen production was published in 2008 [30] where two in-situ-formed ruthenium catalysts were evaluated for hydrogen production under continuous operation conditions. In this case, a closed reactor was used, which also served as hydrogen storage medium. The reaction conditions, such as temperature, amine additive, and [FA:amine] ratio were carefully investigated in order to achieve the optimum performance of the catalytic system at the lowest possible temperatures. Under optimized operation conditions, this system achieved 1,000,000 TONs. Thus, although, to date, the application of the Metal-complex:H_2: HCOOH technology has been demonstrated in only one pilot application, the continuous-operation mode provides encouraging results, in a leap forward towards a scalable in-situ H_2 production technology from HCOOH. Continuous production of hydrogen from FA decomposition has been described in the recent review of Van Putten et al. [67] focusing on scale-up applications, and the review of Durbin and Malardier-Jugroot [32] regarding mainly vehicle applications.

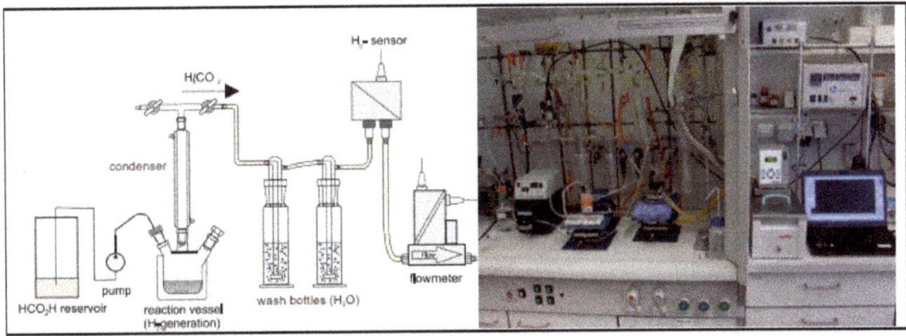

Figure 14. Experimental setup for continuous hydrogen production via FA dehydrogenation. Figure adapted from Reference [43].

3.3. Heterogeneous Catalysts

Catalytic H_2 production via FA dehydrogenation has been shown to be achievable by several types of metal complexes in the homogeneous phase. However, the inherent limitation in separation and reusability of such homogeneous catalysts led the researchers to study heterogeneous catalytic systems. The first heterogeneous systems used for FA dehydrogenation were metal particles—not complexes—[27] mainly operating with FA in gas form, at high temperatures (T > 400 °C) and pressures. Typically, the heterogeneous catalysts used under high temperatures and pressures are metal nanoparticles, metal oxides, and metal nanoparticles supported on different matrices [27]. Liquid phase reactions, able to proceed at near-ambient conditions, have been recently presented in the literature i.e., since 2008. To this front, an alternative approach, exploiting the advantages of homogeneous catalytic metal-complexes, is the heterogenization of the metal complex on a solid matrix. Covalent grafting of the metal complex is required to achieve catalysts stability. Such grafting technology has been exploited for many years by our group for immobilization of oxidation catalysts i.e., Mn complexes [68] and Fe complexes on SiO_2-based [69] or carbonaceous-based supports [70]. However, the grafting technology for H_2 production from FA has been exploited only recently by us [23] (see Figure 15) and others. In 2009, Gan et al. [71] had immobilized a homogeneous Ru[meta-trisulfonatedtriphenylphospine] complex on various matrices i.e., polymers and zeolites. Different immobilization methods i.e., such as ion exchange, polymer immobilization, and adsorption were used. High catalytic activity was achieved in some cases. However, leaching of the catalytically active complex from the surface was a serious drawback i.e., since it drives gradual deactivation of catalyst [71].

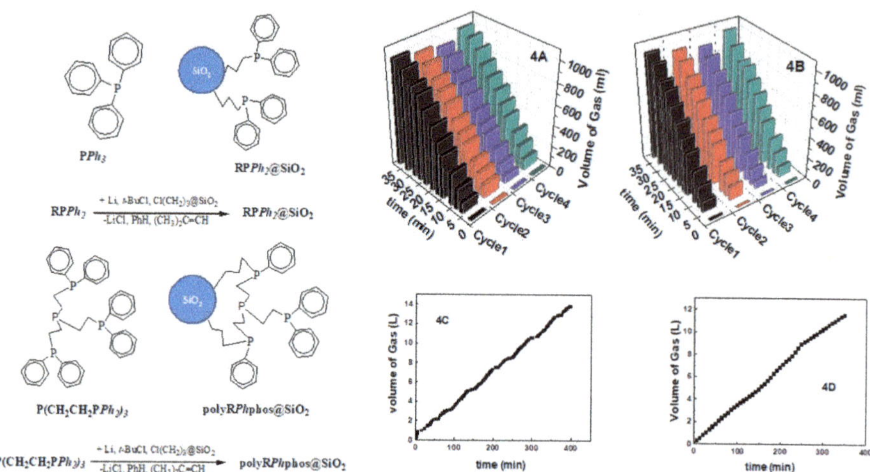

Figure 15. Covalent immobilization of Fe/PPh$_3$ catalyst and catalytic results obtained. Figure adapted from Reference [23].

Table 1. Representative catalytic systems used to FA dehydrogenation.

Catalytic Complex	Temp. °C	Operation Time (h)	TONs	TOFs (h^{-1})	Substrate/Solvent	Additive	FA/Additive Ratio	Ref.
Noble Metals (Homogeneous Catalyst/Homogenous Base Additive)								
Ru$_2$(μ-CO)(CO$_4$)(μ-DPPM)$_2$	RT	-		500	FA/Acetone	-	-	[30]
[Ru(H$_2$O)$_6$](tos)$_2$	120	90	40,000	460	FA, SF/H$_2$O	-	-	[31]
[RuCl$_2$(C$_6$H$_6$)]$_2$/DPPE	40	264	260,000	900	FA/Dimethylamine	-	-	[65]
[Cp*Ir(TMBI)H$_2$O]SO$_4$	80	0.17	10,000	34,000	FA/H$_2$O	-	-	[65]
[Ir(COD)(NP)](tfO)	90	2880	2,160,000	-	FA/SF	-	-	[33]
Cp*Rh(bis-NHC)Cl]Na	100	50	449,000	9000	FA/H$_2$O	-	-	[34]
[RuCl$_2$(p-cymene)]$_2$	40	3	76		FA/H$_2$O	NEt$_3$	2/5	[61]
[RuCl$_2$(p-cymene)]$_2$	40	3	21		FA/H$_2$O	NEt$_3$	3/4	[61]
[RuCl$_2$(p-cymene)]$_2$	40	3	40		FA/H$_2$O	HexNMe$_2$	4/5	[61]
Homogeneous Catalyst/Heterogeneous Base Co-catalyst								
Ru/P(CH$_2$CH$_2$PPh$_2$)	80	696	3924	823	FA/PC	SiO$_2$-NH$_2$	500/1	[26]
Non-Noble metals (Homogeneous catalyst/homogenous additive)								
[Fe(CO)$_{12}$/benzylphospine/TPY	60	51	1266	25	FA/DMF	light	-	[41]
Fe(BF$_4$)$_2$/PPh$_3$	80	18	92,417	5390	FA/PC	-	-	[43]
[(PNP)Fe(H)$_2$(CO)	40	240	100,000	420	FA/THF	Net3	1/2	[44]
[(PNP)Re(CO)$_2$]	180	1	3300	3300	FA/Dioxane		-	[45]
Homogeneous Catalyst/Heterogeneous Base Co-catalyst								
Fe(BF$_4$)$_2$/PPh$_3$	80	200	8483	6245	FA/PC	SiO$_2$	200/1	[54]
Fe(BF$_4$)$_2$/PPh$_3$	80	88	8564	5773	FA/PC	SiO$_2$-COOH	577/1	[62]
Fe(BF$_4$)$_2$/PPh$_3$	80	34	8668	14,942	FA/PC	SiO$_2$-NH$_2$	500/1	[62]
Homogeneous catalyst/homogenous additive: Lewis-Acid Additives								
[PNP$_1$ Fe (H)(CO)(OOCH)]	80	4	231	999	FA/Dioxane	LiBF4	1/0.1	[46]
[PNP1 Fe (H)(CO)(OOCH)]	80	4	263	999	FA/Dioxane	NaCl	1/0.1	[46]
[PNP1 Fe (H)(CO)(OOCH)]	80	4	323	999	FA/Dioxane	NaBF$_4$	1/0.1	[46]
Immobilized catalysts								
Fe(BF$_4$)$_2$-PPh$_3$@SiO$_2$	80	1.25	7869	6295	FA/PC	-		[25]
Pd-P@ SiO$_2$	85	-	719	-	H$_2$O	-	-	[72]
Ru-S-SiO$_2$	85	-	344		H$_2$O	-	-	[72]
Ru-P-SiO$_2$	85	-	102		H$_2$O	-	-	[72]
Pd-N-SiO$_2$	85	-	115		H$_2$O	-	-	[72]

Another example of immobilized catalyst presented by Guo and co-workers [72] consisted of Ru and Pd metal centers with sulfur ligands, covalently attached on SiO_2 support. The best results were obtained for the Pd-S-SiO_2 (TOF = 700 at 85 °C). More recently, heterogenization of an iron-phosphine complex Fe/(tris[(2-diphenylphosphino)ethyl]phosphine) was presented by our group [23]; see Figure 15. Immobilization of the complex on SiO_2 increased the TOFs by 170% vs. the analogous homogeneous system. Importantly, in this work [23], an heterogenized catalyst was demonstrated, for the first time, to have high efficiency and robustness in a wide range of temperatures, being able to produce up to 14 lt of H_2 within 6 h of continuous operation. Thus, the covalent immobilization strategy of the metal-complex or/and the co-catalyst can provide definitive advantages paving the road towards the continuous-operation standards where durability and recyclability will be of primary importance. In 2019, catalytic dehydrogenation of formic acid using molecular Rh and Ir catalysts immobilized on bipyridine-based covalent triazine frameworks was described by Yoon and co-workers [73]. In the case of the Ir catalyst, it achieved TOFs = 7930 h^{-1} i.e., 27 times higher than the corresponding homogeneous Ir Complex.

Oxide-supported metal single-atom or nanoclusters have shown catalytic properties, from the mechanistic point of view the metal nuclearity, and electronic properties can play crucial role in catalytic activity and selectivity [74], for example subnanometer Pt clusters show high selectivity to hydrogenation over dehydration reaction in contrast large Pt NPs. This exemplifies a novel concept i.e., single-atom or sub nanoclusters can be anticipated as catalytic structures between the homogeneous metal-complexes and the nanoparticles.

4. Conclusions

The catalytic mechanism of HCOOH dehydrogenation by molecular (ML) catalysts can be operationally divided in three sequential steps: *(i)* FA activation by deprotonation of one HCOOH molecule and formation of the $HCOO^-$ anion, *(ii)* catalyst activation, *and (iii)* catalytic H_2 production. This process can be accomplished via two alternative routes, conceptually described as Cycle-I and Cycle-II. Cycle-I operates with FA (HCOOH), while Cycle-II with formate anion ($HCOO^-$); both involve a β-hydride transfer from substrate (FA or $HCOO^-$ anion) to metal. The FA activation can be facilitated by organic bases or nanoparticles. Amines in homogeneous phase act as sacrificial additives, while SiO_2 particles or NH_2-funtionalised-SiO_2 nanohybrids can act as co-catalysts. Optimal performance of a catalytic system should consider the role of the solvent as well as the eventual inhibitory role of H_2O. Heterogenization of a ML-catalyst on SiO_2 provides a novel route where the co-catalytic benefit of the silica-OH groups contributes to enhanced performance of the ML catalysts. Another interesting aspect for the catalytic FA dehydrogenation is the acceleration of the catalytic reaction by light. Several experimental issues and limitations, such as the pH of the solution, the process temperature, the stability of the complex, and the additive/co-catalysts should be considered when the long-term continuous operation is envisaged. Minimizing the high concentrations of amine-type additives are a mandatory challenge. Heterogenization of the catalyst and the co-catalyst can serve as multipurpose strategies for long-term continuous operation and economic competitiveness of this technology.

Author Contributions: Writing—original draft preparation P.S., M.S.; writing—review and editing, P.S., M.L., Y.D.; supervision, Y.D., M.L.; project administration, Y.D., M.L.; All authors have read and agreed to the published version of the manuscript.

Funding: This research is co-financed by Greece and the European Union (European Social Fund- ESF) through the Operational Programme «Human Resources Development, Education and Lifelong Learning» in the context of the project "Strengthening Human Resources Research Potential via Doctorate Research" (MIS-5000432), implemented by the State Scholarships Foundation (IKY).

Conflicts of Interest: The authors declare no conflict of interest.

Abbreviations

DPPM	$Ph_2PCH_2PPh_2$
tos	toluene-4-sulfonate
DPPE	1,1-bis(diphenylphosphino)ethane
PNP	2,6-bis(di-*tert*-butylphosphinomethyl)pyridine
CP*	1,2,3,4,5-pentamethylcyclopentadienyl
TMBI	tetramethylbiimidazole
PHEN	1,10-phenanthroline
COD	cyclo-1,5-octadiene
NP	2-(di-*tert*-butylphosphinomethyl)pyridine
NHC	N-heterocyclic carbene
TPY	2,2′:6′,2″-terpyridine
PNP1	$HN(CH_2CH_2(PiPr_2))_2$
PPh3	tris[2-(diphenylphosphino)ethyl]phosphine
TON	Turn over number (TON = mol of produced gasses/mol of catalyst)
TOF	Turn over frequency (TOF = TON/t)

References

1. U.S. Energy Information Administration (EIA). Available online: https://www.eia.gov/ (accessed on 4 December 2019).
2. Mason, J.E. World energy analysis: H_2 now or later? *Energy Policy* **2007**, *35*, 1315–1329. [CrossRef]
3. Markiewicz, M.; Zhang, Y.Q.; Bösmann, A.; Brückner, N.; Thöming, J.; Wasserscheid, P.; Stolte, S. Environmental and health impact assessment of Liquid Organic Hydrogen Carrier (LOHC) systems-challenges and preliminary results. *Energy Environ. Sci.* **2015**, *8*, 1035–1045. [CrossRef]
4. Ogden, J.M. Prospects for building a hydrogen energy infrastructure. *Annu. Rev. Energy Environ.* **1999**, *24*, 227–279. [CrossRef]
5. Holdren, J.P. Energy and sustainability. *Science* **2007**, *315*, 737–738. [CrossRef]
6. Eppinger, J.; Huang, K.W. Formic Acid as a Hydrogen Energy Carrier. *ACS Energy Lett.* **2017**, *2*, 188–195. [CrossRef]
7. Kim, D.; Kim, M. Hydrogenases for biological hydrogen production. *Bioresource Techn.* **2011**, *102*, 8423–8431. [CrossRef]
8. Deligiannakis, Y. Nanomaterials for environmental solar energy technologies: Applications & limitations. *KONA Powder Part. J.* **2018**, *35*, 14–31.
9. Klerke, A.; Christensen, C.H.; Nørskov, J.K.; Vegge, T. Ammonia for hydrogen storage: Challenges and opportunities. *J. Mater. Chem.* **2008**, *18*, 2304–2310. [CrossRef]
10. Jonston, T.C.; Morris, D.J.; Wills, M. Hydrogen generation from formic acid and alcohols using homogeneous catalysts. *Chem. Soc. Rev.* **2010**, *39*, 81–88. [CrossRef]
11. McEvoy, J.P.; Brudvig, G.W. Water-splitting chemistry of photosystem II. *Chem. Rev.* **2006**, *106*, 4455–4483. [CrossRef]
12. Navarro, R.M.; Peña, M.A.; Fierro, J.L.G. Hydrogen production reactions from carbon feedstocks: Fossil fuels and biomass. *Chem. Rev.* **2007**, *107*, 3952–3991. [CrossRef] [PubMed]
13. Kelly, N.A. *Advances in Hydrogen Production, Storage, Distribution*; Wood Head Publishing: Cambrige, UK, 2014.
14. Preuster, P.; Papp, C.; Wasserscheid, P. Liquid organic hydrogen carriers (LOHCs): Toward a hydrogen-free hydrogen economy. *Acc. Chem. Res.* **2017**, *50*, 74–85. [CrossRef] [PubMed]
15. Mellmann, D.; Sponholz, P.; Junge, H.; Beller, M. Formic acid as a hydrogen storage material-development of homogeneous catalysts for selective hydrogen release. *Chem. Soc. Rev.* **2016**, *45*, 3954–3988. [CrossRef] [PubMed]
16. Sang, R.; Kucmierczyk, P.; Dong, K.; Franke, R.; Neumann, H.; Jackstell, R.; Beller, M. Palladium-Catalyzed Selective Generation of CO from Formic Acid for Carbonylation of Alkenes. *J. Am. Chem. Soc.* **2018**, *140*, 5217–5223. [CrossRef]

17. Lv, Q.; Meng, Q.; Liu, W.; Sun, N.; Jiang, K.; Ma, L.; Peng, Z.; Cai, W.; Liu, C.; Ge, J.; et al. Pd–PdO Interface as Active Site for HCOOH Selective Dehydrogenation at Ambient Condition. *J. Phys. Chem. C* **2018**, *122*, 2081–2088. [CrossRef]

18. Marcinkowski, M.D.; Liu, J.; Murphy, C.J.; Liliano, M.L.; Wasio, N.A.; Lucci, F.R.; Flytzani-Stephanopoulos, M.; Sykes, C.H. Selective Formic Acid Dehydrogenation on Pt-Cu Single-Atom Alloys. *ACS Catal.* **2017**, *7*, 413–420. [CrossRef]

19. Sordakis, K.; Tang, C.; Vogt, L.K.; Junge, H.; Dyson, P.J.; Beller, M.; Laurenczy, G. Homogeneous Catalysis for Sustainable Hydrogen Storage in Formic Acid and Alcohols. *Chem. Rev.* **2018**, *118*, 372–433. [CrossRef]

20. Wang, Z.L.; Yan, J.M.; Wang, H.L.; Ping, Y.; Jiang, Q. Pd/C synthesized with citric acid: An efficient catalyst for hydrogen generation from formic acid/sodium formate. *Sci. Rep.* **2012**, *2*, 598. [CrossRef]

21. Singh, A.K.; Singh, S.; Kumar, A. Hydrogen energy future with formic acid: A renewable chemical hydrogen storage system. *Catal. Sci. Technol.* **2016**, *6*, 12–40. [CrossRef]

22. Onishi, N.; Laurenczy, G.; Beller, M.; Himeda, Y. Recent progress for reversible homogeneous catalytic hydrogen storage in formic acid and in methanol. *Coord. Chem. Rev.* **2018**, *373*, 317–332. [CrossRef]

23. Filonenko, G.A.; Van Putten, R.; Schulpen, E.N.; Hensen, E.J.M.; Pidko, E.A. Highly Efficient Reversible Hydrogenation of Carbon Dioxide to Formates Using a Ruthenium PNP-Pincer Catalyst. *Chem. Cat Chem.* **2014**, *6*, 1526–1530.

24. Iglesias, M.; Oro, L.A. Mechanistic Considerations on Homogeneously Catalyzed Formic Acid Dehydrogenation. *Eur. J. Inorg. Chem.* **2018**, *2018*, 2125–2138. [CrossRef]

25. Stathi, P.; Deligiannakis, Y.; Avgouropoulos, G.; Louloudi, M. Efficient H_2 production from formic acid by a supported iron catalyst on silica. *Appl. Catal. A Gen.* **2015**, *498*, 176–184. [CrossRef]

26. Solakidou, M.; Deligiannakis, Y.; Louloudi, M. Heterogeneous amino-functionalized particles boost hydrogen production from Formic Acid by a ruthenium complex. *Int. J. Hydrogen Energy* **2018**, 21386–21397. [CrossRef]

27. Li, Z.; Xu, Q. Metal-Nanoparticle-Catalyzed Hydrogen Generation from Formic Acid. *Acc. Chem. Res.* **2017**, *50*, 1449–1458. [CrossRef] [PubMed]

28. Getoff, N. Photoelectrochemical and photocatalytic methods of hydrogen production: A short review. *Int. J. Hydrogen Energy* **1990**, *15*, 407–417. [CrossRef]

29. Coffey, R.S. The decomposition of formic acid catalysed by soluble metal complexes. *Chem. Commun.* **1967**, 923–924. [CrossRef]

30. Gao, Y.; Kuncheria, J.; Yap, G.P.A.; Puddephatt, R.J. An efficient binuclear catalyst for decomposition of formic acid. *Chem. Commun.* **1998**, 2365–2366. [CrossRef]

31. Fellay, C.; Dyson, P.J.; Laurenczy, G. A viable hydrogen-storage system based on selective formic acid decomposition with a ruthenium catalyst. *Angew. Chem. Int. Ed.* **2008**, *47*, 3966–3968. [CrossRef]

32. Durbin, D.J.; Malardier-Jugroot, C. Review of hydrogen storage techniques for on board vehicle applications. *Int. J. Hydrogen Energy* **2013**, *38*, 14595–14617. [CrossRef]

33. Celaje, J.J.A.; Lu, Z.; Kedzie, E.A.; Terrile, N.J.; Lo, J.N.; Williams, T.J. A prolific catalyst for dehydrogenation of neat formic acid. *Nat. Commun.* **2016**, *7*, 11308. [CrossRef] [PubMed]

34. Jantke, D.; Pardatscher, L.; Drees, M.; Cokoja, M.; Herrmann, W.A.; Kuhn, F.E. Hydrogen Production and Storage on a Formic Acid/Bicarbonate Platform Using Water-Soluble N-Heterocyclic Carbene Complexes of Late Transition Metals. *ChemSusChem* **2016**, *9*, 284. [CrossRef] [PubMed]

35. Iguchi, M.; Onishi, N.; Himeda, Y.; Kawanami, H. Ligand Effect on the Stability of Water-Soluble Iridium Catalysts for High-Pressure Hydrogen Gas Production by Dehydrogenation of Formic Acid. *ChemPhysChem* **2019**, *20*, 1296–1300. [CrossRef] [PubMed]

36. Gao, Y.; Kuncheria, J.K.; Jcnkins, H.A.; Puddephatt, R.J.; Payap, G. The interconversion of formic acid and hydrogen/carbon dioxide using a binuclear ruthenium complex catalyst. *J. Chem. Soc. Dalt. Trans.* **2000**, 3212–3217. [CrossRef]

37. Loges, B.; Boddien, A.; Junge, H.; Beller, M. Controlled generation of hydrogen from formic acid amine adducts at room temperature and application in H_2/O_2 fuel cells. *Angew. Chem. Int. Ed.* **2008**, *47*, 3962–3965. [CrossRef] [PubMed]

38. Himeda, Y. Highly efficient hydrogen evolution by decomposition of formic acid using an iridium catalyst with 4,4′-dihydroxy-2,2′-bipyridine. *Green Chem.* **2009**, *11*, 2018–2022. [CrossRef]

39. Fukuzumi, S.; Kobayashi, T.; Suenobu, T. Unusually large tunneling effect on highly efficient generation of hydrogen and hydrogen isotopes in pH-selective decomposition of formic acid catalyzed by a heterodinuclear iridium-ruthenium complex in water. *J. Am. Chem. Soc.* **2010**, *132*, 1496–1497. [CrossRef]

40. Filonenko, G.A.; Van Putten, R.; Hensen, E.J.M.; Pidko, E.A. Catalytic (de)hydrogenation promoted by non-precious metals-Co, Fe and Mn: Recent advances in an emerging field. *Chem. Soc. Rev.* **2018**, *47*, 1459–1483. [CrossRef]

41. Boddien, A.; Loges, B.; Garthner, F.; Torborg, C.; Fumino, K.; Junge, H.; Ludwig, R.; Beller, M. Iron-catalyzed Hydrogen production from formic acid. *J. Am. Chem. Soc.* **2010**, *132*, 8924–8934. [CrossRef]

42. Scotti, N.; Psaro, R.; Ravasio, N.; Zaccheria, F. A new Cu-based system for formic acid dehydrogenation. *RSC Adv.* **2014**, *4*, 61514–61517. [CrossRef]

43. Boddien, A.; Mellmann, D.; Gärtner, F.; Jackstell, R.; Junge, H.; Dyson, P.J.; Laurenczy, G.; Ludwig, R.; Beller, M. Efficient dehydrogenation of formic acid using an iron catalyst. *Science* **2011**, *333*, 1733–1736. [CrossRef] [PubMed]

44. Zell, T.; Butschke, B.; Ben-David, Y.; Milstein, D. Efficient hydrogen liberation from formic acid catalyzed by a well-defined iron pincer complex under mild conditions. *Chem. Eur. J.* **2013**, *19*, 8068–8072. [CrossRef] [PubMed]

45. Vogt, M.; Nerush, A.; Diskin-Posner, Y.; Ben-David, Y.; Milstein, D. Reversible CO2 binding triggered by metal-ligand cooperation in a rhenium(i) PNP pincer-type complex and the reaction with dihydrogen. *Chem. Sci.* **2014**, *5*, 2043–2051. [CrossRef]

46. Zhang, Y.; Würtele, C.; Lagaditis, P.O.; Bielinski, E.A.; Hazari, N.; Schneider, S.; Bernskoetter, W.H.; Mercado, B.Q. Lewis Acid-Assisted Formic Acid Dehydrogenation Using a Pincer-Supported Iron Catalyst. *J. Am. Chem. Soc.* **2014**, *136*, 10234–10237.

47. Mellone, I.; Gorgas, N.; Bertini, F.; Peruzzini, M.; Kirchner, K.; Gonsalvi, L. Selective Formic Acid Dehydrogenation Catalyzed by Fe-PNP Pincer Complexes Based on the 2,6-Diaminopyridine Scaffold. *Organometallics* **2016**, *35*, 3344–3349. [CrossRef]

48. Curley, J.B.; Smith, N.E.; Bernskoetter, W.H.; Hazari, N.; Mercado, B.Q. Catalytic Formic Acid Dehydrogenation and CO2 Hydrogenation Using Iron PNRP Pincer Complexes with Isonitrile Ligands. *Organometallics* **2018**, *37*, 3846–3853. [CrossRef]

49. Fellay, C.; Yan, N.; Dyson, P.J.; Laurenczy, G. Selective formic acid decomposition for high-pressure hydrogen generation: A mechanistic study. *Chem. Eur. J.* **2009**, *15*, 3752–3760. [CrossRef]

50. Stathi, P.; Mitrikas, G.; Sanakis, Y.; Louloudi, M.; Deligiannakis, Y. Back-clocking of Fe^{2+}/Fe^{1+} spin states in a H_2-producing catalyst by advanced EPR. *Mol. Phys.* **2013**, *111*, 18–19. [CrossRef]

51. Mellmann, D.; Barsch, E.; Bauer, M.; Grabow, K.; Boddien, A.; Kammer, A.; Sponholz, P.; Bentrup, U.; Jackstell, R.; Junge, H.; et al. Base-free non-noble-metal-catalyzed hydrogen generation from formic acid: Scope and mechanistic insights. *Chem. Eur. J.* **2014**, *20*, 13589–13602. [CrossRef]

52. Yang, X. Mechanistic insights into iron catalyzed dehydrogenation of formic acid: β-hydride elimination vs. direct hydride transfer. *Dalt. Trans.* **2013**, *42*, 11987–11991. [CrossRef]

53. Sánchez-De-Armas, R.; Xue, L.; Ahlquist, M.S.G. One site is enough: A theoretical investigation of iron-catalyzed dehydrogenation of formic acid. *Chem. Eur. J.* **2013**, *19*, 11869–11873. [CrossRef] [PubMed]

54. Stathi, P.; Deligiannakis, Y.; Louloudi, M. Co-catalytic enhancement of H_2 production by SiO_2 nanoparticles. *Catal. Today* **2015**, *242*, 146–152. [CrossRef]

55. Stathi, P.; Louloudi, M.; Deligiannakis, Y. Efficient Low-Temperature H_2 Production from HCOOH/HCOO– by [Pd^0@SiO_2-Gallic Acid] Nanohybrids: Catalysis and the Underlying Thermodynamics and Mechanism. *Energy Fuels* **2016**, *30*, 8613–8622. [CrossRef]

56. Wang, W.-H.; Xu, S.; Manaka, Y.; Suna, Y.; Kambayashi, H.; Muckerman, J.T.; Fujita, E.; Himeda, Y. Formic Acid Dehydrogenation with Bioinspired Iridium Complexes: A Kinetic Isotope Effect Study and Mechanistic Insight. *ChemSusChem* **2014**, *7*, 1976–1983. [CrossRef] [PubMed]

57. Manca, G.; Mellone, I.; Bertini, F.; Peruzzini, M.; Rosi, L.; Mellmann, D.; Junge, H.; Beller, M.; Ienco, A.; Gonsalvi, L. Inner-versus outer-sphere ru-catalyzed formic acid dehydrogenation: A computational study. *Organometallics* **2013**, *32*, 7053–7064. [CrossRef]

58. Linn, D.E.; King, R.B.; King, A.D. Catalytic reactions of formate: Part 1. Photocatalytic hydrogen production from formate with chromium hexacarbonyl. *J. Mol. Catal.* **1993**, *80*, 151–163. [CrossRef]

59. Onishi, M. Decomposition of formic acid catalyzed by hydrido (phosphonite) cobalt (I) under photoirradiation. *J. Mol. Catal.* **1993**, *80*, 145–149. [CrossRef]

60. Loges, B.; Boddien, A.; Junge, H.; Noyes, J.R.; Baumann, W.; Beller, M. Hydrogen generation: Catalytic acceleration and control by light. *Chem. Commun.* **2009**, 4185–4187. [CrossRef]

61. Junge, H.; Boddien, A.; Capitta, F.; Loges, B.; Noyes, J.R.; Gladiali, S.; Beller, M. Improved hydrogen generation from formic acid. *Tetrahedron Lett.* **2009**, *50*, 1603–1606. [CrossRef]

62. Stathi, P.; Deligiannakis, Y.; Louloudi, M. Co-catalytic Effect of Functionalized SiO_2 Material on H_2 Production from Formic Acid by an Iron Catalyst. *Mater. Res. Soc. Symp. Proc.* **2014**, 1641. [CrossRef]

63. Papp, G.; Ölveti, G.; Horváth, H.; Kathó, A.; Joó, F. Highly efficient dehydrogenation of formic acid in aqueous solution catalysed by an easily available water-soluble iridium(III) dihydride. *Dalt. Trans.* **2016**, *45*, 14516–14519. [CrossRef] [PubMed]

64. Boddien, A.; Loges, B.; Junge, H.; Gärtner, F.; Noyes, J.R.; Beller, M. Continuous hydrogen generation from formic acid: Highly active and stable ruthenium catalysts. *Adv. Synth. Catal.* **2009**, *351*, 2517–2520. [CrossRef]

65. Morris, D.J.; Clarkson, G.J.; Wills, M. Insights into hydrogen generation from formic acid using ruthenium complexes. *Organometallics* **2009**, *28*, 4133–4140. [CrossRef]

66. Fujita, E.; Muckerman, J.T.; Himeda, Y. Interconversion of CO_2 and formic acid by bio-inspired Ir complexes with pendent bases. *Biochim. Biophys. Acta Bioenerg.* **2013**, *1827*, 1031–1038. [CrossRef]

67. Van Putten, R.; Wissink, T.; Swinkels, T.; Pidko, E.A. Fuelling the hydrogen economy: Scale-up of an integrated formic acid-to-power system. *Int. J. Hydrogen Energy* **2019**, *44*, 28533–28541. [CrossRef]

68. Serafimidou, A.; Stamatis, A.; Louloudi, M. Manganese(II) complexes of imidazole based-acetamide as homogeneous and heterogenised catalysts for alkene epoxidation with H_2O_2. *Catal. Commun.* **2008**, *9*, 35–39. [CrossRef]

69. Bilis, G.; Stathi, P.; Mavrogiorgou, A.; Deligiannakis, Y.; Louloudi, M. Improved robustness of heterogeneous Fe-non-heme oxidation catalysts: A catalytic and EPR study. *Appl. Catal. A Gen.* **2014**, *470*, 376–389. [CrossRef]

70. Mavrogiorgou, A.; Papastergiou, M.; Deligiannakis, Y.; Louloudi, M. Activated carbon functionalized with Mn(II) Schiff base complexes as efficient alkene oxidation catalysts: Solid support matters. *J. Mol. Catal. A Chem.* **2014**, *393*, 8–17. [CrossRef]

71. Gan, W.; Dyson, P.J.; Laurenczy, G. Heterogeneous silica-supported ruthenium phosphine catalysts for selective formic acid decomposition. *ChemCatChem* **2013**, *5*, 3124–3130. [CrossRef]

72. Zhao, Y.; Deng, L.; Tang, S.Y.; Lai, D.M.; Liao, B.; Fu, Y.; Guo, Q.X. Selective decomposition of formic acid over immobilized catalysts. *Energy Fuels* **2011**, *25*, 3693–3697. [CrossRef]

73. Gunasekar, G.H.; Kim, H.; Yoon, S. Dehydrogenation of formic acid using molecular Rh and Ir catalysts immobilized on bipyridine-based covalent triazine frameworks. *Sustain. Energy Fuels* **2019**, *3*, 1042–1047. [CrossRef]

74. Kuo, C.; Lu, Y.; Kovarik, L.; Engelhard, M.H.; Karim, A.M. Structure Sensitivity of Acetylene Semi-Hydrogenation on Pt Single atoms and Subnanometer clusters. *ACS Catal.* **2019**, *9*, 11030–11041. [CrossRef]

MDPI

St. Alban-Anlage 66

4052 Basel

Switzerland

Tel. +41 61 683 77 34

Fax +41 61 302 89 18

www.mdpi.com

Energies Editorial Office

E-mail: energies@mdpi.com

www.mdpi.com/journal/energies